Organic Photovoltaics
and Related Electronics—
From Excitons to Devices

T0331238

MATERIALS RESEARCH SOCIETY
SYMPOSIUM PROCEEDINGS VOLUME 1270

Organic Photovoltaics and Related Electronics— From Excitons to Devices

Symposium held April 5–9, 2010, San Francisco, California, U.S.A.

EDITORS:

Venkat Bommisetty
South Dakota State University

Garry Rumbles
National Renewable Energy Laboratory

Sean E. Shaheen
University of Denver

Peter Peumans
Stanford University

Niyazi Serdar Sariciftci
Johannes Kepler University of Linz

Jao van de Lagemaat
National Renewable Energy Laboratory

K.S. Narayan
Jawaharlal Nehru Centre
for Advanced Scientific Research

Gilles Dennler
Konarka Technologies, Inc.

Symposium II Organizers:

Zhenan Bao
Stanford University

Alejandro L. Briseno
University of Massachusetts

Vitaly Podzorov
Rutgers University

Iain McCulloch
Imperial College London

Materials Research Society
Warrendale, Pennsylvania

CAMBRIDGE
UNIVERSITY PRESS

University Printing House, Cambridge CB2 8BS, United Kingdom

One Liberty Plaza, 20th Floor, New York, NY 10006, USA

477 Williamstown Road, Port Melbourne, VIC 3207, Australia

314-321, 3rd Floor, Plot 3, Splendor Forum, Jasola District Centre, New Delhi - 110025, India

103 Penang Road, #05-06/07, Visioncrest Commercial, Singapore 238467

Cambridge University Press is part of the University of Cambridge.

It furthers the University's mission by disseminating knowledge in the pursuit of
education, learning and research at the highest international levels of excellence.

www.cambridge.org
Information on this title: www.cambridge.org/9781605112473

Materials Research Society
506 Keystone Drive, Warrendale, PA 15086
http://www.mrs.org

First published 2010
First paperback edition 2012

Single article reprints from this publication are available through
University Microfilms Inc., 300 North Zeeb Road, Ann Arbor, MI 48106

CODEN: MRSPDH

A catalogue record for this publication is available from the British Library

ISBN 978-1-605-11247-3 Hardback
ISBN 978-1-107-40672-8 Paperback

CONTENTS

MORPHOLOGY

POSTER SESSION II

DEVICES II

CHARGE TRANSPORT

POSTER SESSION: ORGANIC MATERIALS FOR OFETs, OLEDs, AND OPV

INTERFACES AND MORPHOLOGY

PREFACE

Symposium GG, "Nanoscale Charge Transport in Excitonic Solar Cells," Symposium HH, "Organic Photovoltaic Science and Technology," and Symposium II, "Materials Science and Charge Transport in Organic Electronics," were held April 5–9 at the 2010 MRS Spring Meeting in San Francisco, California. These three symposia were coordinated to create an organic electronic "super symposium." A total of 516 abstracts were received between the three, illustrating the tremendous interest in the field and amount of research activity that is being undertaken today. Among several joint sessions that were held, a "super session" on Advancing Organic Photovoltaics was assembled that consisted of a full day of invited talks from Alan Heeger, Kees Hummelen, René Janssen, Mike McGehee, Jean-Luc Bredas, Mark Thompson, Karl Leo, and Yang Yang. The room, filled to capacity, was treated to a series of exciting new results, illustrations of fundamental concepts, and also some historical perspectives on the field. A capstone event during the week was a lively and very interactive panel discussion, moderated by Serdar Saracifti, which featured all of the jointly invited speakers from the week.

Carrying out research in the field of organic electronics requires, at least, a working knowledge of a wide variety of disciplines—from physics, to chemistry, to materials science, to optical and electrical engineering. This is evident in the panoply of topics that appeared in the conference. This proceedings volume presents a snapshot of the presentations and discussions that took place, covering areas such as: the synthesis of new materials that show increasing control of nanostructured properties; the design and implementation of new characterization techniques that probe local molecular structure or exciton and charge carrier dynamics; studies of interfaces; techniques for optical and electrical modeling of devices; and optimization of processing methods for improved device performance.

Organic electronics is now established as a technology suitable for some consumer electronic applications, as evidenced by a variety of commercially available devices. Whether or not these will have staying power in the face of heated competition from inorganic technologies remains to be seen. However, there is good reason to believe that organic electronics are here to stay, at least for applications that can take advantage of their inherent properties. Longer-term questions are: What broader societal impact can the field have? Can it contribute substantively to a new renewable energy-based economy? Can it provide new types of electronic circuitry that enable ubiquitous computing that is novel in form and function? To reach these lofty goals will require continued progress at all levels in the field, from fundamental science to engineering to implementation. We hope that this MRS conference and proceedings are able to provide one more little push toward these endeavors.

Sean E. Shaheen
Venkat Bommisetty

November 2010

MATERIALS RESEARCH SOCIETY SYMPOSIUM PROCEEDINGS

MATERIALS RESEARCH SOCIETY SYMPOSIUM PROCEEDINGS

Prior Materials Research Society Symposium Proceedings available by contacting Materials Research Society

Poster Session

Mater. Res. Soc. Symp. Proc. Vol. 1270 © 2010 Materials Research Society   1270-GG04-05

Analysis of Charge Transport and Recombination Studied by Electrochemical Impedance Spectroscopy for Dye-Sensitized Solar Cells with Atomic Layer Deposited Metal Oxide Treatment on TiO₂ Surface

Braden Bills[*], Mariyappan Shanmugam, Mahdi Farrokh Baroughi and David Galipeau
Department of Electrical Engineering and Computer Science
South Dakota State University, Brookings, SD 57007, USA

ABSTRACT

The performance of dye-sensitized solar cells (DSSCs) is limited by the back-reaction of photogenerated electrons from the porous titanium oxide (TiO₂) nanoparticles back into the electrolyte solution, which occurs almost exclusively through the interface. This and the fact that DSSCs have a very large interfacial area makes their performance greatly dependant on the density and activity of TiO₂ surface states. Thus, effectively engineering the TiO₂/dye/electrolyte interface to reduce carrier losses is critically important for improving the photovoltaic performance of the solar cell. Atomic layer deposition (ALD), which uses high purity gas precursors that can rapidly diffuse through the porous network, was used to grow a conformal and controllable aluminum oxide (Al_2O_3) and hafnium oxide (HfO_2) ultra thin layer on the TiO₂ surface. The effects of this interfacial treatment on the DSSC performance was studied with dark and illuminated current-voltage and electrochemical impedance spectroscopy (EIS) measurements.

INTRODUCTION

DSSCs have been under development for almost two decades and are showing great potential to compete with conventional p-n junction solar cells in terms of efficiency, cost and simple, fast, low-temperature and non-vacuum fabrication procedures. However, major challenges still persist including narrow band absorption dyes, poor stability, lack of robustness in large scale production and the corrosive nature of the liquid electrolyte. Another significant problem limiting the performance of DSSCs is photogenerated carrier loss that occurs between the photoelectrode and the electrolyte [1]. Carrier loss is fundamentally different in DSSCs compared to conventional solar cells. Recombination between electrons and holes in conventional solar cells occurs mainly in the bulk of usually only one material (absorber), where recombination in DSSCs is due to electron injection from semiconducting nanoparticles to liquid electrolyte occurring almost exclusively through the interface. This and the fact that DSSCs have a much larger (~1000 times) interfacial area compared to conventional solar cells makes their performance greatly dependant on the density and activity of surface states. Thus, effectively engineering the interfaces to reduce carrier losses is critically important for improving open-circuit voltage (V_{OC}), short-circuit current (J_{SC}), fill factor (FF) and efficiency (η).

Recent works have shown DSSC performance enhancements when a wide range of wet chemical processed materials were deposited onto the mesoporous TiO₂ network [2]. Another technique to treat surface defects is to use oxygen, nitrogen, hydrogen or argon plasma to passivate TiO₂ sub-dioxides such as Ti_2O_3 and TiO resulting in slightly enhanced DSSC performance compared to wet chemical processed treatments [3]. A new approach for depositing an interfacial layer onto the mesoporous TiO₂ network is by using ALD, which is advantageous over wet chemical method for treating defective interfaces with metal oxides due to the higher

---

[*] Corresponding author: E-mail: Braden.Bills@sdstate.edu

purity of gas precursors, faster diffusion of gas precursors into the mesoporous network and better conformallity and controllability of metal oxide thickness [4-6].

In our previous preliminary work, we showed that ALD Al_2O_3 and HfO_2 grown on the surface of mesoporous TiO_2 will improve the photovoltaic performance of DSSCs, and the role of the interfacial layer thickness (5, 10 and 20 cycles) was studied [6]. More importantly, however, is to understand the effects that the metal oxide surface treatment has on the physical processes at the interface, which is critical for continual improvements in the reduction of photogenerated electron loss at the interface, subsequently improving overall solar cell efficiency. In this work, DSSCs with 20 cycles of ALD Al_2O_3 and 5 cycles of ALD HfO_2, which showed the greatest enhancement (E) in efficiency for each material, were further studied using EIS, where the impedance spectra was fitted to an equivalent circuit model and the extracted charge transport, accumulation and transfer (recombination) parameters were used to interpret the DSSC performance. Dark and illuminated current-voltage (I-V) measurements were also used for characterizing the metal oxide surface treatments, and were related to the EIS parameters.

EXPERIMENTAL DETAILS

The DSSC photoelectrode was prepared by RF-sputtering a 50 nm thick TiO_2 compact layer on indium tin oxide (ITO) coated glass substrate at 150 W RF-power and 20 sccm argon flow rate and annealed for 30 min. at 450 °C. TiO_2 colloidal paste (Solaronix) was applied onto the compact layer and sintered for 30 min. at 450 °C resulting in a thickness of ~6 μm. Next, interfacial metal oxide treatment was deposited onto the mesoporous TiO_2, where the untreated photoelectrode will be referred to as the reference. Al_2O_3 and HfO_2 ultra thin (a few atomic layers) layers of 20 and 5 cycles respectively were deposited onto the mesoporous network by ALD (Savannah 100, Cambridge NanoTech). Tri methyl aluminum and hafnium tetra chloride gases and water vapor (H_2O) were used as the Al, Hf and oxygen precursors respectively and were sequentially applied into the deposition chamber at 200 °C. This process is very short, where each cycle was ~10 seconds. The three photoelectrodes were sensitized with N719 dye. The counter electrode was prepared by RF-sputtering a 40 nm thick Pt layer on ITO coated glass substrate. A spacer was placed between the two electrodes and filled with Iodolyte AN-50 (I^-/I_3^-) electrolyte.

I-V measurements were performed with an Agilent 4155C semiconductor parameter analyzer and illumination was obtained by an AM1.5 filtered xenon arc lamp at 100 mW/cm^2. A NREL calibrated Hamamatsu S1133 photodetector, which absorbs photons in 350-800 nm wavelength similar to N719 based DSSC, was used to calibrate the power at 100 mW/cm^2. EIS measurements were performed with a computer controlled HP/Agilent 4192A LF impedance analyzer. All EIS measurements were done in the dark with AC signal amplitude of 10 mV, frequency range of $5-10^5$ Hz, and applied forward bias of 0–1V.

RESULTS AND DISCUSSION

DSSCs with metal oxide treatment showed a significant change in dark I-V behavior as the dark saturation currents were suppressed as shown in the low to medium forward bias regions of Figure 1a. This suggests that the electronic quality, in terms of density and activity of surface states, of the mesoporous TiO_2/dye/electrolyte interface had been improved with metal oxide treatment. Also, electron transport was not blocked by the metal oxide layer as shown in the high forward bias region of Figure 1a. The small electron affinities of Al_2O_3 and HfO_2 (1.35 and 2 eV

4

respectively) compared to that of TiO$_2$ (3.9eV) suggest that photoexcited electrons can be injected from the dye to the TiO$_2$ by tunneling through the ultra thin metal oxide interfacial layer. The illumination characteristics of Table 1 and Figure 1b show that J$_{SC}$ increases compared to the reference for both metal oxides and V$_{OC}$ improves with HfO$_2$ treatment. Possible reasons for enhancement of solar cell performance with metal oxide surface treatment will be further investigated with EIS measurements.

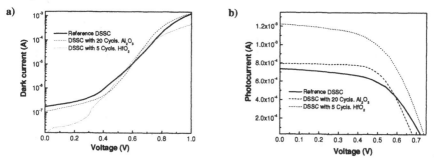

Figure 1. (a) Dark and (b) illuminated I-V characteristics of the reference, DSSC with 20 cycles of Al$_2$O$_3$ and DSSC with 5 cycles of HfO$_2$.

Table 1. Illuminated solar cell parameters of reference and DSSCs with 20 and 5 cycles of Al$_2$O$_3$ and HfO$_2$ treatments.

Type of Photoelectrode	J$_{SC}$ (mA/cm^2)	V$_{OC}$ (mV)	FF	η (%)	E (%)
TiO$_2$	10.5	730	0.55	4.2	--
TiO$_2$/ Al$_2$O$_3$ (20 Cyc.)	11.4	684	0.64	5.0	19
TiO$_2$ / HfO$_2$ (5 Cyc.)	17.5	745	0.55	7.1	69

The shape of experimental EIS spectra is explained extensively elsewhere [7]; however, a few key points will be mentioned. At low forward biases, TiO$_2$ is non-conducting and charge transfer only occurs at the TCO/compact layer/electrolyte interface. At high forward bias, TiO$_2$ becomes conductive and the TiO$_2$/dye/electrolyte or TiO$_2$/metal oxide/dye/electrolyte becomes the dominant interface due to its much large interfacial area.

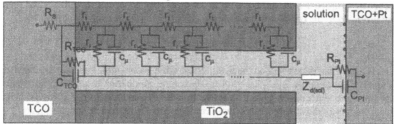

Figure 2. Equivalent circuit model of a DSSC developed by others [7-9].

Experimental EIS data was fitted to a previously developed equivalent circuit model shown in Figure 2 [7-9]. Three important parameters from the equivalent circuit model for characterizing interfacial layers are the transport resistance (R_t), the capacitance of the film (C_{film}) and the interfacial charge transfer (recombination) resistance (R_{ct}). Figure 3 shows the semilogrithmic plot of R_t, C_{film}, R_{ct} and the electron lifetime (τ_n) for the reference and DSSCs with 20 and 5 cycles of Al_2O_3 and HfO_2 interfacial treatment respectively.

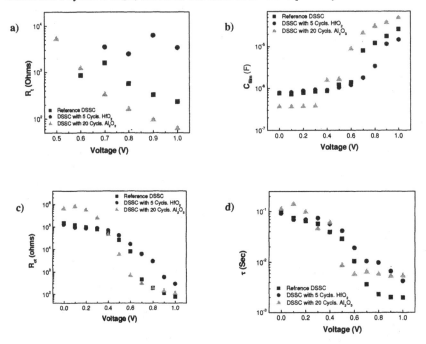

Figure 3. Fitting results of (a) R_t, (b) C_{film}, (c) R_{ct} and (d) τ_n of the reference (black squares), DSSC with 20 cycles of Al_2O_3 (green triangles) and DSSC with 5 cycles of HfO_2 (red circles).

Transport Resistance Figure 3a shows an exponential dependence of R_t with applied forward bias and is related to the conduction band energy level by the expression [7]

$$R_t = R_0 \exp\left[\frac{-E_{Fn} - E_{cb}}{kT}\right],\tag{1}$$

where R_0 is a constant depending on the geometrical dimensions of the cell, E_{Fn} the quasi-Fermi energy level, E_{cb} the energy level of the lower edge of the conduction band, k the Boltzmann constant and T the temperature. Figure 3a shows that the R_t trend for HfO_2 and Al_2O_3 surface treatments are larger and smaller respectively compared to the reference, suggesting a displacement of the conduction band energy level [7]. Larger R_t values corresponded to an upward shift in E_{cb}, which agrees with the respective increase and decrease of V_{OC} for HfO_2 and Al_2O_3 treated DSSCs compared to the reference. The lower R_t for Al_2O_3 showed that electron transport improved and was not blocked at the interface agreeing with the high dark current density in the high bias region of Figure 1a. Similarly, the higher R_t and lower dark current for HfO$_2$ showed that electron transport was reduced.

Film Capacitance Figure 3b shows C_{film} increased exponentially once the TiO$_2$ became conductive which agreed with the behavior of R_t; that is, they both exhibited exponential behavior at the same bias for a given surface treatment. C_{film} is the sum of capacitances at the TCO/compact layer/electrolyte interface and the chemical capacitance, C_μ, at the TiO$_2$/dye/electrolyte or TiO$_2$/metal oxide/dye/electrolyte interface expressed by [8,9]

$$C_\mu = \frac{e^2}{kT} \exp\left[\frac{\alpha}{kT}(E_{Fn} - E_{cb})\right], \qquad (2)$$

where e is the electron charge and α is a constant related to the distribution of electronic states below the conduction band. The density of accessible electronic states within the TiO$_2$ nanoparticles or at the TiO$_2$/dye/electrolyte or TiO$_2$/metal oxide/dye/electrolyte interface corresponded to capacitance [8]. The decreased C_{film} of HfO$_2$ treatment compared to the reference may be explained in two ways: 1) as the conduction band was displaced upward, the chance of an electron in the conduction band getting trapped by an accessible electronic state became less probable due to the increased displacement in energy levels and 2) the density of easily accessible electronic states were reduced which agrees with the suppressed low bias dark currents of Figure 1a. The increased C_{film} of Al$_2$O$_3$ treatment compared to the reference in the exponential region was most likely due to the downward displacement of the conduction band making electron capture by accessible electronic states more likely to occur.

Charge Transfer (Recombination) Resistance The overall charge transfer resistance, R_{ct}, was composed of the parallel combination of charge transfer resistances at the TCO/compact layer/electrolyte and TiO$_2$/dye/electrolyte or TiO$_2$/metal oxide/dye/electrolyte interfaces. Three regions of R_{ct} are shown in Figure 3c. At the low forward bias region, TiO$_2$ behaved as an insulator and electron recombination occurred at the TCO/compact layer/electrolyte interface. When TiO$_2$ became conductive, the slope of R_{ct} changed with an exponential dependence on forward bias signifying that the majority of recombination occurred at the TiO$_2$/dye/electrolyte or TiO$_2$/metal oxide/dye/electrolyte interface. At the high forward bias region, the slope of R_{ct} decreased due to the series resistance of the cell being the same order in magnitude as R_{ct} [8]. In the exponential region at intermediate biases, R_{ct} can be expressed by [8,9]

$$R_{ct} = R_0 \exp\left[-\frac{\beta}{kT}(E_{Fn} - E_{redox})\right], \qquad (3)$$

where β is the transfer coefficient, E_{redox} the energy level of the redox couple and $E_{Fn} - E_{redox} = eV_a$ where V_a is the applied bias. Applying eq. 3 to the exponential region of Figure 3c yields β values of 0.397, 0.511 and 0.242 for the reference, DSSC with 20 cycles of Al$_2$O$_3$ and DSSC with 5 cycles of HfO$_2$ respectively. The larger β value for 20 cycles of Al$_2$O$_3$ treatment coincided with the largest fill factor value. Recombination of electrons directly from the conduction band to the electrolyte requires a value of $\beta \approx 1$; since this is not the case for these DSSCs, transfer of electrons must occur through surface states before recombining with electrolyte [9]. Larger R_{ct} values with 5 cycles of HfO$_2$ treatment in the exponential region resulted in reduced charge losses thus contributing to V_{OC} improvement; likewise, smaller R_{ct} values with 20 cycles of Al$_2$O$_3$ in the exponential region resulted in increased charge losses contributing to worsened V_{OC} compared to the reference. Similar to the R_t and C_{film} analysis described above, larger R_{ct} values support that electrons in the conduction band are less likely to get captured by surface states either due to an increase in the conduction band energy level or reduced density of easily accessible electronic states or both.

Electron Lifetime The τ_n is the product of R_{ct} and C_μ and is in direct relation to J_{SC} and the overall DSSC performance as shown in Figure 3d. At middle and high bias conditions, τ_n is largest for 5 cycles of HfO_2 treatment, corresponding to the highest J_{SC} and conversion efficiency enhancements. The τ_n for 20 cycles of Al_2O_3 treatment improved compared to the reference for all bias conditions except at intermediate biases agreeing with the moderate J_{SC} and conversion efficiency enhancements.

CONCLUSIONS

Electrochemical impedance spectroscopy measurements can provide insight into performance enhancement of DSSCs treated with atomic layer deposited ultra thin metal oxide interfacial layers, between the photoelectrode and electrolyte, by the study of the charge transport, accumulation and transfer (recombination) parameters. These impedance parameters strongly agree with dark and illuminated current-voltage measurements. The largest improvement in DSSC performance was observed by DSSCs with 5 cycles of HfO_2 interfacial treatment, where the performance enhancement was attributed to a reduced density of easily accessible electronic states and a low recombination rate of electrons in TiO_2. DSSCs with 20 cycles of Al_2O_3 had a marginal increase in performance.

ACKNOWLEDGEMENTS

Authors would like to thank American Science and Technology, NASA South Dakota Space Grant Consortium-#NNG05GJ98H and NSF-EPSCoR-0554609 program for partial funding of this work and the Nanofabrication Center of the University of Minnesota for providing the ALD facility.

REFERENCES

1. Junghanel, M.; Tributsch, H, C. R. Chimie 2006, 9, 652.
2. Sommeling, P. M.; O'Regan, B. C.; Haswell, R.; Smit, H. J. P.; Bakker, N. J.; Smits, J. J. T.; Kroon, J. M.; Van Roosmalen, J. A. M. J. Phys. Chem. B 2006, 110, 19191.
3. Kim, Y.; Yoo, B. J.; Vittal, R.; Lee, Y.; Park, N. G.; Kim, K. J. J. Power Sources 2008, 175, 914.
4. T. C. Tien, F. M. Pan, L. P. Wang, C. H. Lee, Y. L. Tung, S. Y. Tsai, C. Lin, F. Y. Tsai and S. J. Chen, Nanotechnology 2009, 20, 305201
5. T. W. Hamann, O. K. Farha and J. T. Hupp, J. Phys. Chem. C 2008, 112, 19756.
6. M. Shanmugam, M. F. Baroughi, D. Galipeau, Thin Solid Films 2009 doi:10.1016/j.tsf.2009.08.033.
7. Fabregat-Santiago, F.; Bisquert, J.; Garcia-Belmonte, G.; Boschloo, G.; Hagfeldt, A. Sol. Energy Mater. Sol. Cells 2005, 87,117.
8. Fabregat-Santiago, F.; Bisquert, J.; Palomares, E.; Otero, L.; Kuang, D. B.; Zakeeruddin, S. M.; Grätzel, M. J. Phys. Chem. C 2007,111, 6550.
9. Wang, Q.; Ito, S.; Grätzel, M.; Fabregat-Santiago, F.; Mora-Seró, I.; Bisquert, J.; Bessho, T.; Imai, H. J. Phys. Chem. B 2006, 110, 25210.

Mater. Res. Soc. Symp. Proc. Vol. 1270 © 2010 Materials Research Society   1270-GG04-09

Photoconductivity Measurements of Organic Polymer/Nanostructure Blends

David Black and Shashi Paul[1]
Emerging Technologies Research Centre, Hawthorn Building, DeMontfort University, The Gateway, Leicester, England, LE1 9BH

ABSTRACT

In an attempt to produce low cost and high quality polymer/nanoparticle blends for use in hybrid organic/inorganic photovoltaic devices we prepared blends of dihexylsexithiophene and tetragonal barium titanate particles. These polymer nanoparticle blends were deposited as films by spin coating and sublimation. The films were characterised and compared using a wide range of techniques; The electrical photoconductivity analysis of these structures carried out using an HP4140B picoammeter and a solar simulator after aluminium gap cell electrodes had been deposited on the films by sublimation, spectroscopic studies (UV-VIS) were carried out to understand the photoconductivity measurements and ellipsometry was used to determine the thickness of the films. The photoconductivity of the spin coated films was the highest reaching 8.5×10^{-10}A at 20 V, the sublimed films reached ~4 x 10^{-10}A at 40V. This is thought to be due to the thinness of the sublimed films combined with the inhomogeneous distribution of nanoparticles compared with the spin coated film. Sublimed films have been shown by others to be better structured than spin coated films, if this property can be utilized with further optimization of the sublimation process then this technique offers the potential to produce very thin high quality films for use in organic and hybrid photovoltaic devices.

INTRODUCTION

Hybrid organic/inorganic photovoltaic devices comprising a blend or blends of polymers and inorganic nanostructures offer the potential of flexible devices with a small number of steps in the fabrication process. These devices are excitonic in nature, when an organic photoconductive material absorbs a photon of an appropriate wavelength, an excited state is created. It has been demonstrated by other groups such as that of Roman and other that the efficiency of photovoltaic devices is increased by incorporating nano-particulates into the organic photoconductive materials [1].

This work follows on from our previous work using tetragonal barium titanate to increase the permittivity of both insulating and conducting polymers with the aim of producing nanoparticle polymer matrices suitable for inclusion in hybrid organic/inorganic photovoltaic devices [2]. Commercially available barium titanate (BT) nanoparticles are cubic in nature and do not have the ferroelectric characteristics needed to increase the permittivity of polymers and hence we used tetragonal BT, prepared as described below, which does have the required properties. It has also been demonstrated previously by Kim et al [3] that BT is generally not highly soluble and by attaching a suitable ligand it can be more easily made into a suspension for spin coating.

An initial study has been undertaken in our lab to understand photoconductivity of the blends of dihexylsexithiophene (DH6T) and tetragonal barium titanate particles. Here we have examined the possibility of using thermal evaporation, also known as sublimation, to create thin

---

[1] Contact author email: spaul@dmu.ac.uk

films of DH6T containing barium titanate nanoparticles with and without ligands. These films were compared against a sample film of DH6T with barium titanate (BT) nanoparticles with an n-octyl phosphonic acid ligand (OPA) prepared by spin coating.

EXPERIMENT

A series of samples were prepared including one spin coated control sample and the blends of polymer and nanoparticles for sublimation detailed in Table 1.

Table 1 – Polymer-nanoparticle blends used in sublimated films

Sample	DH-6T (mg)	BT (mg)	BT/OPA(mg)	Ave. Film Thickness (nm)
0	20	0	0	63
1	20	20	0	66
2	20	0	20	78

Control sample

A series of samples were prepared including spin coated control sample and the blends of polymer and nanoparticles for sublimation detailed in Table 1. The spin coated polymer blend contained 20mg/ml DH6T solution in dichlorobenzene with 20mg/ml BT/OPA. The structure of the samples used for electrical measurements is shown in Figure 1. The ferroelectric BT particles were prepared from cubic BT nanoparticles that had been annealed in air at 1000°C for 1 hr in a similar way to that described by Kwon and Yoon[4]. The OPA ligand was then attached using the methodology suggested by Kim et al [3].

A small amount of solution was taken from each sample vial and spin coated at 3000 rpm using a Specialty Coating Systems model G3P-8 spin coater onto a pre-prepared slide (1mm thick 25mm x 25mm Corning glass slides as substrate) immediately after being taken from the ultrasonic bath. A second sample was prepared on a p-type silicon substrate. Once all the samples were spin coated they were allowed to fully dry in air. The glass substrate was moved to a thermal evaporator used only for depositing to metal contacts. The top Al contacts were evaporated at a pressure of ~ 1 x 10^{-6} mbar through a gap cell shadow mask.

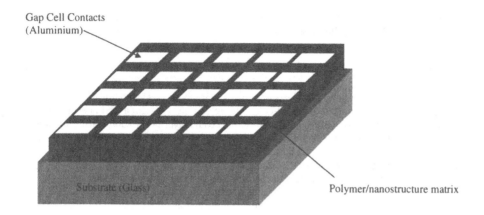

Gap Cell Contacts
(Aluminium)

Substrate (Glass)

Polymer/nanostructure matrix

Figure 1 – Typical sample gap cell structure

Sublimated samples

The sublimated samples were prepared by mixing dry polymer with nanoparticles with and without the OPA ligand in the quantities detailed in Table 1. Each sample powder was then placed on a clean tungsten boat and placed in the thermal evaporator intended for use only with polymers and nanoparticles. The evaporator was left to pump down and achieved a final pressure of ~ 1 x 10^{-6}mbar. The powder was then sublimated onto glass and p-type silicon substrates. This process was repeated for each blend of polymer and nanoparticle.

The glass substrates were then moved to the metal evaporator and gap cell electrodes were deposited as described above. The device architecture is as shown in Figure 1.

Testing

The thin films on the p-type silicon substrate were used to determine thickness and composition using an ellipsometer with an accuracy of ±1nm. UV-Vis measurements were carried out on samples deposited on corning glass substrates, the Thermo Scientific Evolution 300 spectrometer used in these experiments takes a background scan of clean substrate prior to testing the thin films on the same substrate and uses a clean reference substrate in the scanning process. This allows the substrate to be corrected for automatically during the scanning process and produces a spectra that is only for the film and not the substrate.

Each of the completed samples had a structure as depicted in Figure for electrical measurements. The samples were then taken for electrical testing, in a solar simulator chamber using an AM1.5 solar source for illumination. The voltage was varied between 0 and 20V for the spin coated sample and 0 and 40V for the sublimed samples and the current was measured under illuminated and non illuminated conditions as the voltage increased and decreased.

RESULTS AND DISCUSSION

Quality and uniformity of films is of paramount importance when fabricating organic electronic devices, this work show that it is possible to deposit polymer/nanoparticle blends by sublimation that are comparable to good quality spin coated layers. The sublimated films were both similar in thickness being ~63nm ±1 while the spin coated film was slightly thicker at 78nm ±1. A visual inspection of the films showed that the spin coated material was a noticeably darker shade of pink/orange than the sublimated samples.

The UV-Vis spectroscopy reveals that both the spin coated and sublimated samples contain absorption peaks at ~275nm and ~360nm, which correspond to energies of 4.5 and 3.44eV respectively. Duhm et al [5] postulate that these energy levels represent a group of 6 narrowly spaced localized π-states, which would be excited by the absorption of photons in this wavelength, potentially leading to the release of Frenkel singlet excitons as suggested by Horowitz et al [6].

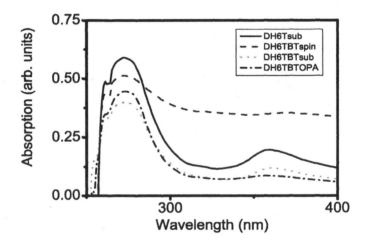

Figure 2 UV-Vis absorption spectrum for sublimated samples of the blends used in this work, dihexyl-sexithiophene (DH6T), dihexyl-sexithiophene and barium titanate (DH6TBT) and dihexyl-sexithiophene and barium titanate with n-octyl phosphonic acid ligand (DH6TBTOPA).

Figure 3 below details the current-voltage (IV) characteristics of three different films, fig.-3a shows the characteristics of the spin coated sample, while 3b and 3c show the characteristics of sublimed DH6T films with BT and BT/OPA respectively. The DH6T film was

12

omitted as there was no appreciable difference between the illuminated and dark states. As can be seen the spin coated sample has the highest photoconductivity (8.5x 10^{-10}A) and reaches this at only 20V whereas both sublimated samples achieve comparable photoconductivity at approximately twice that voltage. It can be seen that the photoconductivity of all of the samples in dark conditions are comparable, although again the spin coated film is higher than that of the sublimated films.

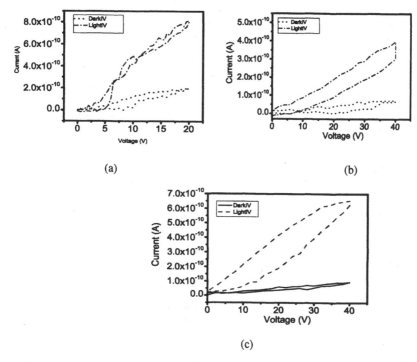

(a)

(b)

(c)

Figure 3 - IV characteristics of selected films; 3a – spin coated DH6T with BT/OPA, 3b – sublimated DH6T with BT, 3c – sublimated DH6T with BT/OPA.

There are two potential reasons for the higher photoconductivity in the spin coated films as opposed to the sublimated films, firstly the spin coated films are thicker than the sublimated films which may account for the generally higher absorption in the visible spectrum, secondly and most importantly the density of nanoparticles per unit area is higher in the spin coated sample as these materials were created from a homogeneous solution containing equal amounts of polymer and nanoparticles by mass. The sublimated samples were prepared using dry powders of equal mass, but due to the potential differences in evaporation rates it was not possible, at this time, to determine a methodology to ensure that the ratio was maintained in the sublimated film. AFM images, which there is not shown here, revealed that the nanoparticles in the sublimated films were quite widely spaced unlike the spin coated materials. However, it should be noted that sublimation offers the ability to orient the polymer chains, relative to

13

particular substrates, in such a way as to optimize their electrical properties as described by Duhm et al [5].

CONCLUSIONS

This work demonstrates that it is possible to obtain photoconductive polymer/nanoparticle films by sublimation. While these films are not currently as efficient as spin coated films this is most likely due to the thickness of the films and the distribution of the nanoparticles within the polymer matrix. Further optimization of the sublimation process to determine the optimum current at which both materials can be sublimated in as uniform a manner as possible should be investigated, this combined with additional methods for improving film quality such as; heating the substrate during sublimation and annealing the film after deposition.

Sublimation offers the potential to simply and cheaply produce good quality photoconductive films that could be incorporated into hybrid organic/inorganic photovoltaic devices.

REFERENCES

1. L. S. Roman, in *Organic Photovoltaics: Mechanisms, Materials and Devices*, edited by S.S Sun and N.S. Sariciftci (Taylor and Francis, Boca Raton, 2005), Vol. 1, pp. 367-386.
2. D. Black, S. Paul, and I. Salaoru, "Ferro-electric Nanoparticles in Polyvinyl Acetate (PVAc) Matrix - A Method to Enhance the Dielectric Constant of Polymers," Journal of Nanoscience and Nanotechnology, In Press, (2010).
3. P. Kim, S. C. Jones, P. J. Hotchkiss et al., "Phosphonic acid-modified barium titanate polymer nanocomposites with high permittivity and dielectric strength," Advanced Materials **19** (7), 1001-1005 (2007).
4. Sung-Wook Kwon and Dang-Hyok Yoon, "Tetragonality of nano-sized barium titanate powder prepared with growth inhibitors upon heat treatment," Journal of the European Ceramic Society **27** (1), 247-252 (2007).
5. S. Duhm, I. Salzmann, N. Koch et al., "Vacuum sublimed dihexylsexithiophene thin films: Correlating electronic structure and molecular orientation," Journal of Applied Physics **104**, 033717 (2008).
6. H. Glowatzki, S. Duhm, K. F. Braun et al., "Molecular chains and carpets of sexithiophenes on Au(111)," Physical Review B **76** (12), 125425 (2007).

Mater. Res. Soc. Symp. Proc. Vol. 1270 © 2010 Materials Research Society 1270-GG04-14

Charge Carrier Transport in the Bulk and at the Surface of Nanoparticles: a Quasi-Solid-State Dye-Sensitized Solar Cell

Dennis Friedrich[1] and Marinus Kunst[1]

[1]Helmholtz-Zentrum Berlin für Materialien und Energie, Institute Solar Fuels, Hahn-Meitner Platz 1, 14109 Berlin, Germany

ABSTRACT

A quasi-solid-state dye-sensitized solar cell is presented, where the conventional liquid electrolyte is replaced by an electrolyte film, reaching a solar light-to-current conversion efficiency of 3%. Contactless transient photoconductance measurements were performed, revealing decay behavior of photoinduced charge carriers, dependent on external applied potential conditions. The measurements show that the decay is controlled by the injection of electrons into the front contact, hindered or enhanced by the field in the space charge region.

INTRODUCTION

A solar cell is presented, which avoids some of the disadvantages of the conventional liquid dye-sensitized solar cell (DSSC) [1]. This cell consists also of sensitized TiO_2 nanoparticles but the electrolyte is reduced to a stable film on the surface of the nanoparticles. It still includes a redox pair (Iodine-Iodide), specific chemicals and also strongly bound H_2O molecules. This cell has still a modest efficiency but it promises improved stability and a significant potential for further optimization. Fortunately, the absence of a liquid electrolyte in this cell allows the application of highly performing transient techniques to the complete cell such as transient photoconductivity in the microwave frequency range (TRMC). Knowledge on kinetic charge transfer behavior is not only important for the improvement of this cell but also for the understanding of charge transport in nanosystems in general.

EXPERIMENTAL DETAILS

Materials

All solvents and reagents, unless otherwise stated, were of puriss. quality and used as received. Transparent conductive glass, TEC 7 (FTO, sheet resistance 7 Ω/sq, thickness 2.3 mm) was purchased from Pilkington. TiO_2 Paste DSL 18NR-AO and Dye N719 from Dyesol were used as received. Ethanol, acetonitrile (anhydrous, 99.8%), 2-propanol, titanium (IV) chloride tetrahydrofuran complex (1 : 2) ($TiCl_4$*2THF, 97%), 4-t-butylpyridine (99%) and iodine (99.99%, metals basis) were provided by Sigma-Aldrich. Lithium iodide (anhydrous) and t-butanol were purchased from Fluka. Printex XE2 carbon black was provided by Degussa-Evonik. Milli-Q (Millipore) grade water was used in all experiments.

Device Fabrication

The front electrode was prepared by screen printing TiO_2 layers of 1 cm^2 area on FTO substrates, resulting in a 12 μm thick layer of TiO_2 particles. The substrates were sintered at 520 °C for 30 min, cooled, followed by a post-treatment in a 50 mM $TiCl_4$ solution for 30 min at 70 °C, flushed with water and again sintered for 30 min at 450 °C [2]. After the second heat treatment the substrates were cooled to 60 °C, immersed into the dye solution (1 mM of N719 in acetonitrile/t-butanol), and stored at room temperature for ~16 h. After sensitization, 3 μL of the electrolyte solution (5 M lithium iodide/0.05 M iodine/0.05 M 4-t-butylpyridine in ethanol) were deposited on the electrode. After 24 h, a layer of carbon nanoparticles was deposited on the substrate by spraying a carbon suspension (Printex XE2 in 2-propanol) with an airbrush, acting as catalyst for the reduction of iodine and as the electric contact to the counter electrode. The electrodes were stored for another 24 h to remove volatile compounds from the carbon layer. Finally, a FTO glass was pressed on top of the front electrode, fixed with epoxy resin, and the cell was contacted.

I-V Measurements

The I-V characteristics of the solar cells were determined by using a solar simulator (VOSS Electronic GmbH, WXS-140S-Super). AM1.5G (equivalent to an irradiance of 100 mW cm^{-2}) illumination conditions were simulated by combined use of a 100 W Xenon lamp and a 120 W halogen lamp. Measurements were performed both at irradiation intensity of 100 mW cm^{-2} and 10 mW cm^{-2}, while the latter was realized by the use of a neutral density filter. The system was periodically calibrated with a Si-reference solar cell. The current-voltage characteristics were obtained by applying an external potential bias to the cell and by measuring the generated photocurrent with a measuring unit (Keithley SMU 238). The temperature of the sample holder was set constant by a thermostat at a given value, normally set at 23 °C for standard measurements.

Time Resolved Microwave Conductivity Measurements (TRMC)

Transient photoconductance measurements in the microwave frequency range were performed by the Time Resolved Microwave Conductivity (TRMC) technique in a Ka-band (28.5 – 40 GHz) apparatus as described previously [3,4]. The excitation occurred by 10 ns (FWHM) pulses of a Nd-YAG laser at a wavelength of 532 nm with a diameter of about 3 mm. The excitation intensity was adjusted by the use of calibrated filters. External potentials were applied by connecting the front contact of the cell with the working electrode, and the back contact to the combined counter- and reference electrode of the potentiostat (Wenking POS 73). The potentials were set at 0.1 V steps for the regime from 0.6 to -0.6 V.

The TRMC signal, $\Delta P(t)/P$, is the relative change of the microwave power reflected by the sample induced by a photogenerated change of the conductance ($\Delta S(t)$) [3]. In general the TRMC signal is determined by all mobile excess species but for the case of TiO_2 it will be assumed that only electrons at the bottom of the conduction band with mobility μ_e contribute to the photoconductance:

$$\frac{\Delta P(t)}{P} = A\Delta S(t) = Ae\Delta N(t)\mu_e \tag{1}$$

A is a sensitivity factor depending on the experimental configuration and the electrical parameters of the sample [4]. The total number of excess electrons with mobility μ_e at time t, $\Delta N(t)$, refers to an integration of the excess electron concentration, $\Delta n(t)$, only over the thickness d of the sample. Any process that decreases the number of excess electrons in the conduction band leads to a decay of the photoconductivity.

Photoconductance measurements by the TRMC technique allow the characterization of nanoporous systems such as DSSC, providing information on charge carrier kinetics for injection as well as recombination processes. Under external applied potential conditions, the measured signal is only sensitive to the photoinduced change of the conductance. This allows the analysis of charge carrier kinetics independent of external measurement conditions.

DISCUSSION

The I-V curve of our quasi solid-state dye-sensitized solar cell is shown in figure 1. Reasonable values for Open-Circuit Potential and fill factors could be obtained, whereas the rather small current seems to be the limiting parameter. The cell has a better performance for reduced light intensities. This indicates a higher carrier loss with increasing excess charge concentration. The most probable reason for this effect is a saturation of the transport of the positive charge from the excited dye to the counter electrode. Therefore, the recombination reaction of injected electrons, e_{TiO2}, with oxidized redox species R^+, as well as the recombination of electrons with the oxidized dye may limit the short-circuit photocurrent density. The overall smaller photocurrent, compared to conventional liquid iodine/iodide dye cells, may be due to a incomplete covering of the sensitized TiO_2 with the electrolyte film, partially preventing the regeneration of the excited dye.

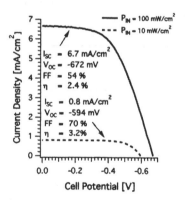

Figure 1. Photo-I-V characteristic of a quasi solid-state dye-sensitized solar cell under AM1.5G illumination conditions (100 mW/cm^2, solid line) and 10 mW/cm^2 (dashed line).

The results of the TRMC measurements for different applied external potentials are presented in a double-logarithmic plot in figure 2. The TRMC signals show a potential dependence. The decay is rather slow in the short time range, up to a few microseconds. Much slower than the previously reported power-law decay behavior of excess charge carriers in dye sensitized TiO_2 films, usually ascribed to a dispersive transport of electrons in the TiO_2 film [5]. In our case the rather high excitation densities may induce a saturation of states in the conduction band tail, leading to the slow initial decay.

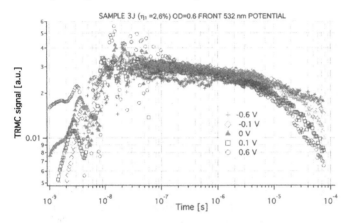

Figure 2. Double-logarithmic plot of the TRMC signal from a quasi solid-state dye-sensitized solar cell induced by 532 nm light pulses at an excitation density of 20 mJ/cm^2. Signals are plotted for an applied external potential to the cell as indicated.

Starting from about 3 µs, the TRMC signal shows a decay of carriers and a dependence on the applied potential, which can be better observed in a semi-logarithmic plot as shown in figure 3. Here, the TRMC signal is presented together with the results of the exponential fits for two time intervals. The TRMC signal reveals an increase of the decay time with increasing external potential; reaching the maximum at 0.6 V and the minimum at -0.6 V.

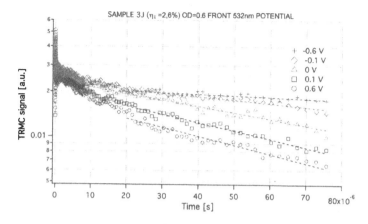

Figure 3. Semi-logarithmic plot of the TRMC signal from a quasi sold state dye-sensitized solar cell induced by 532 nm light pulses at an excitation density of 20 mJ/cm². Signals are plotted for an applied external potential to the cell as indicated. Included are the exponential Fits of the decay times for time intervals 200 ns to 8µs (solid lines) and 20 µs to 75 µs (dashed lines).

Figure 4 represents the lifetimes for the two time intervals as a function of the applied external potential. The lifetime of the photogenerated charge carriers increases with rising negative applied potential. There are in principle three decay channels for the electron injected from the dye into the TiO_2: (1) The injected electron can recombine with the oxidized dye, (2) with the oxidized redox species in the electrolyte and (3) as well be injected into the front contact. Regarding the potential dependence of the decay we believe it to be mainly depending on the injection of charges into the front contact because the predominant influence of an applied external potential is at the FTO/TiO_2 interface. Although there is a noticeable change for the "short" regime (from ~ 20 µs at 0.6 V to ~ 80 µs at -0.6 V), the increase of the carrier lifetime is considerably stronger in the "long" time regime (from ~ 75 µs at 0.6 V to ~ 450 µs at -0.6 V). The lifetime reaches its maximum at an external potential of -0.6 V. This potential lies near the Open-Circuit potential of the cell. The photocurrent density in this potential regime is minimal, i.e. the conclusion can be drawn that the potential dependent decay observed is limited by the injection of electrons into the front contact. The field in the space charge region at the FTO/TiO_2 interface controls this injection. The onset of the decay at about 3 µs can be interpreted as the time required for the excess charge carriers to reach the front contact and to enable electron injection into the FTO contact.

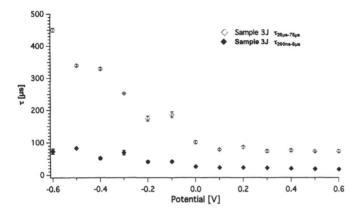

Figure 4. Lifetime τ of the photogenerated charge carriers as a function of the applied external potential. Solid diamonds (open diamonds) represent the lifetime from the exponential fit of the TRMC signal for the time regime of 200 ns to 8 μs (20 μs to 75 μs).

CONCLUSIONS

A quasi solid-state dye sensitized solar cell is presented with an efficiency of 3%. Excess charge carrier decay was shown to be dependent on external applied potential conditions. Electron injection into the front contact limits this decay and is controlled by the space charge field.

ACKNOWLEDGMENTS

The authors would like to thank the Deutsche Bundesstiftung Umwelt (DBU) for financial support.

REFERENCES

1. B. O'Regan and M. Grätzel, Nature **353**, 737-740 (1991).
2. P.M. Sommeling, B.C. O'Regan, R.R. Haswell, H.J.P. Smit, N.J. Bakker, J.J.T. Smits, J.M. Kroon, and J.A.M. van Roosmalen, J. Phys. Chem. B **110**, 19191-19197 (2006).
3. M. Kunst and G. Beck, J. Appl. Phys. **60**, 3558-3566 (1986).
4. C. Swiatkowski, A. Sanders, K. Buhre, and M. Kunst, J. Appl. Phys. **78**, 1763-1775 (1995).
5. M. Kunst, F. Goubard, C. Colbeau-Justin, and F. Wünsch, Mater. Sci. Eng., C **27**, 1061-1064 (2007).

Mater. Res. Soc. Symp. Proc. Vol. 1270 © 2010 Materials Research Society   1270-GG04-16

Impacts of ZnO/TiO$_2$ Assorted Electrode on Photoelectric Characteristics of Dye-Sensitized Solar Cells

Xu Wang, Haiyou Yin, Bao Wang, Lifeng Liu, Yi Wang, Jinfeng Kang[*]
Institute of Microelectronics, Peking University, Beijing 100871, P.R. China
[*]Email: kangjf@pku.edu.cn

ABSTRACT

A novel ZnO/TiO$_2$ assorted photoelectrode for dye-sensitized solar cells (DSSCs) is proposed. The impacts of the ZnO/TiO$_2$ assorted photoelectrode on the photovoltaic performance of dye-sensitized solar cells (DSSCs) were investigated. The measurements of the light transmission spectra showed the higher transmittance through ZnO/FTO than through FTO during the effective wavelength region of 536nm~800nm for DSSCs, indicating that ZnO/TiO$_2$ assorted photoelectrode is beneficial for the photovoltaic performance of DSSCs. The measurements on the photovoltaic characteristics of the DSSC cell indicate that the inserted ZnO layer can cause the increased open circuit voltage (V_{oc}) more than 70 mV and fill factor (FF) but the decreased short circuit current. The enhanced V_{oc} and FF could be attributed to the suppressed the recombination of photon-generated carriers between the ZnO/TiO$_2$ assorted photoelectrode and electrolyte (dye) compared to TiO$_2$ photoelectrode. However, the additional series resistance of inserted ZnO layer causes the reduced short circuit current. The optimized conversion efficiency can be achieved in the DSSC with ZnO/TiO$_2$ assorted photoelectrode by using low series resistance of ZnO layer.

INTRODUCTION

Dye-sensitized solar cells (DSSCs) have been widely studied as a promising solar cell technology because of its low-cost and environmental-friendship [1, 2]. However, some critical issues such as the relatively low conversion efficiency and large-area integration are still block the practical application of DSSCs even though various technical solutions were proposed to solve the bottleneck problems [3-6].

It is well known that electron mobility of ZnO is much higher than that of TiO$_2$, while the conduction band edge of both materials is approximate at the same level [3, 4]. Therefore ZnO was expected to be a good candidate as electron acceptor and transport material in DSSCs [5]. However, the reported photocurrent conversion efficiency of ZnO DSSCs is far less than that of TiO$_2$ due to the poor dye absorption of ZnO [6]. In this paper, we proposed a novel ZnO/TiO$_2$ assorted photoelectrode to improve the performance of DSSCs. The inserted ZnO layer between TiO$_2$ and FTO is anticipated to enhance the electron transport efficiency to FTO due to the higher electron mobility than TiO$_2$ and the blocking effect on the direct contact between FTO and electrolyte. The measured data demonstrated that the new ZnO/TiO$_2$ assorted photoelectrode could effectively improve fill factor (FF) and the open circuit voltage V_{oc} but the reduced series resistance of the inserted ZnO layer is needed to achieve the higher short circuit current.

EXPERIMENT

In this study, ZnO and Ga-doped ZnO were fabricated by sol-gel process. For the ZnO, the 5g two hydrated zinc acetate [Zn(CH COO)$_2$ • 2H$_2$O] was mixed in 30mL methyl alcohol [CH$_3$OH] 2.2mL diethanolamine, then the mixed liquid was fully stirred in the surrounding of 60 ℃ water bath until forming the sol. Next the ZnO layer was firstly deposited on FTO by spray-coating by 3000r/min for 20s using a rotary coating machine, followed by a furnace annealing treatment at 500℃. For the Ga-doing ZnO, Gallium doping was realized by mixing a saturated solution (C=0.3M) of gallium acetylacetonate [Ga(C$_5$H$_7$O$_2$)$_3$, abridged as GaAcAc], in acetylacetone [(C$_5$H$_8$O$_2$, abridged as AcAc] with the sol of zinc acetate[7]. Then the TiO$_2$ paste was screen printed on ZnO or Ga-doped ZnO layer to form the 1×1cm^2 assorted layer photoelectrode. The samples with the assorted photoelectrode was baked in air at a temperature of 450℃ for 2h. After that, the samples were immersed in a 0.5mM N719 dye solution adsorbing dye at room temperature for 20h. Finally the redox electrolyte was introduced into the assembled DSSC by using the Surlyn film (25 um) as the sealing spacer between the front and back plates.. A schematic diagram of the DSSC is shown in Figure 1.

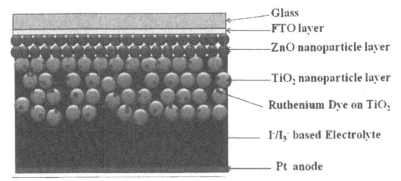

Figure 1. The schematic diagram of the DSSC with the assorted electrode.

The photovoltaic performances of the DSSCs were measured by an I-V measurement system under a light intensity of 100 mW/cm^2 (1 sun). Ga-doped ZnO layer was also deposited to study the impact of the inserted ZnO layer conductivity on the DSSC performance.

RESULTS AND DISCUSSION

Figure 2 shows the scanning electron microscopy (SEM) image and X-ray diffraction spectrum of ZnO layer deposited by sol-gel process on FTO followed by a furnace annealing at 500°C in air ambience. The film thickness is typically 100-800nm. The crystallized ZnO films with the average size being about 50~80nm were deposited.

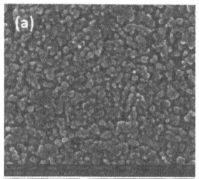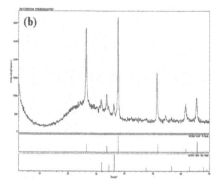

Figure 2. Structural characteristics of the deposited ZnO nanoparticle layer on FTO. a) SEM image, and b)X-ray diffraction spectrum.

The optimal TiO_2 layer thickness of about (12 ± 1) um screen printed on ZnO layer for the DSSCs was measured.

Figure 3 shows the typically measured current -voltage curves of DSSCs with and without the ZnO nanoparticle layer. The increased V_{OC} more than 70 mV and higher fill factor (FF) are observed in the DSSC samples with ZnO layer compared to the counterparts without ZnO layer while the lower conversion efficiency was measured in the DSSC samples with ZnO layer due to the lower short circuit current (J_{SC}).

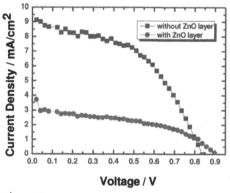

Figure 3. Measured I-V curves of DSSCs with and without a ZnO layer.

Figure 4 shows the dependences of short circuit current density and open voltage on the thickness of the inserted ZnO layer. The increased J_{sc} and almost constant V_{oc} were observed with the reduced ZnO layer's thickness, implying that the reduced short circuit current density is mainly due to additional series resistance of the ZnO layer and the optimized DSSCs can be realized by reducing the additional series resistance of the inserted ZnO layer.

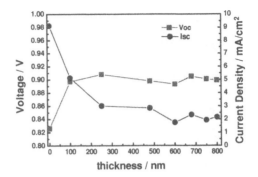

Figure 4. Dependence of short circuit current density and open voltage as a function of thickness of the inserted ZnO layer.

Figure 5 shows the measured transmittance of FTO with and without the ZnO nanoparticle layer. The higher transmittance was measured in the ZnO/FTO sample between the wavelength region of 536nm and 800nm, which indicates that the additional ZnO enhances the transmittance of the effective light spectra for DSSCs. The transmittance can be optimized in the thinner ZnO layer samples.

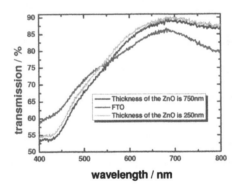

Figure 5. The light transmission spectrum through FTO with and without the ZnO nanoparticle layer.

Figure 6 shows dark current of DSSC samples with and without the ZnO layer. The reduced dark current was observed in the DSSCs with ZnO layer compare to that without ZnO layer. This indicates that the inserted ZnO layer can effectively enhance the transport of photon-generated carriers from the electrode to FTO layer and suppress the carriers' recombination between electrodes and electrolyte (dye). The enhanced transport and suppressed recombination cause the increased open circuit voltage and the improved fill factor.

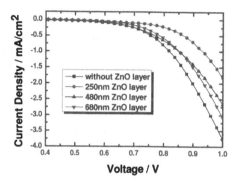

Figure 6. I-V characteristics in dark measured in DSSCs with and without a ZnO layer.

Based on above observations, we conclude that a thinner and lower resistance of inserted ZnO layer is effective to improve the conversion efficiency of DSSCs. In order to identify this conclusion, Ga-doped inserted ZnO layer with lower resistivity than pure ZnO was applied. The doping density of Ga into ZnO is about 5%. The measured I-V characteristics of DSSCs are shown in Figure 7, which clearly show that the higher conversion efficiency can be achieved in the DSSC samples with the reduced series resistance of the inserted ZnO layer corresponding to the thinner thickness and reduced resistivity.

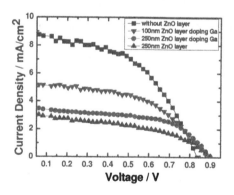

Figure 7. I-V characteristics of DSSCs with and without Ga doping.

CONCLUSIONS

In this study, a novel ZnO/TiO_2 assorted electrode is proposed to optimize the photovoltaic performance of DSSCs. The results indicate that the inserted ZnO nanoparticle layer can effectively enhance the transmission of incidence light in the effective light spectra region of DSSCs and the photon-generated carriers transport, suppress the recombination of photon-generated carriers and the dark current. This will benefit to increase the open circuit voltage and the fill factor, but the additional serial resistance of inserted ZnO layer is a critical factor for the conversion efficiency, which may cause the significant reduction of the short circuit current. The improved conversion efficiency can be achieved by using thin thickness and low resistivity of ZnO layer with Ga doping. Our results demonstrated that Ga doped ZnO/TiO_2 assorted electrode is promising for DSSCs application. Overall, the photovoltaic performance of DSSCs can be optimized by using a thinner and lower resistivity ZnO layer. Our study can provide a new technical solution for the optimization of the DSSCs.

REFERENCES

1. O'Regan B., Grätzel M., *Nature*, **353**, 737–739(1991).
2. Grätzel M., *Journal Photochemistry and Photobiology A: Chemistry*, **164**, 23–14(2004).
3. Marı́a Quintana, Tomas Edvinsson, Anders Hagfeldt, and Gerrit Boschloo, *J. Phys. Chem. C*, **111**, 1035-1041(2007).
4. Seok-Soon Kim, Jun-Ho Yum, Yung-Eun Sung, *Journal of Photochemistry and Photobiology A: Chemistry*, **171**, 269–273(2005).
5. S. Sakthivel, B. Neppolian, M.V. Shankar, B. Arabindoo, M. Palanichamy, V. Murugesan, *Solar Energy Materials & Solar Cells*, **77**, 65-82(2003).
6. Yinhua Jiang, Min Wu, Xiaojuan Wu, Yueming Sun, Hengbo Yin, *Materials Letters*, **63**, 275-278 (2009).
7. A. Tiburcio-Silver, A. Sanchez-Juarez, A. Avila-Garcia, *Solar Energy Materials and Solar Cells*, **55**, 3-10(1998).

Charge Transport

Mater. Res. Soc. Symp. Proc. Vol. 1270 © 2010 Materials Research Society   1270-GG10-02

Using bulk heterojunction field effect measurements to understand charge transport in solar cell materials

Christopher J. Lombardo and Ananth Dodabalapur
Microelectronics Research Center, The University of Texas at Austin, 10100 Burnet Rd., Bldg 160, Austin, TX 78758, U.S.A.

ABSTRACT

Ambipolar organic thin-film transistors (OTFTs) have been used to study the transport of charge carriers in bulk heterojunction (BHJ) organic photovoltaic devices. Active layers of phase separated blend of poly(3-hexylthiophene) (P3HT) and [6,6]-phenyl C_{61}-butyric acid methyl ester (PCBM), have been chosen due to their use in performance BHJ organic photovoltaic devices as well as ease of device fabrication. A method for determining recombination rate after exciton dissociation and measurement of excess carrier lifetime has been reported by studying drain current behavior which yields carrier mobility, conductivity, and carrier concentration both in dark and AM1.5g illumination. Channel-length dependent measurements of the photocurrent show that significant recombination of separated charge carriers begins to occur at lengths greater than 10 μm. A recombination rate of 2.6×10^{19} cm^{-3} s^{-1} and a carrier lifetime of ≥ 8.8 ms has been calculated.

INTRODUCTION

Organic photovoltaic (OPV) cells have been actively studied for over 25 years and during that time, power conversion efficiency has increased from about 1% to about 8% [1, 2], yet much more needs to be known about the movement of charge carriers within OPV cells and the nature of recombination. Some researchers have developed transient methods like CELIV [3] and photo CELIV [4], but these methods are typically time consuming and require specialized equipment, making them not suitable for OPV manufacturers. We employ ambipolar OTFTs to study charge transport in BHJ PV cells as many researchers are familiar with fabricating and measuring FET parameters. Ambipolar OTFTs have been studied by multiple researchers in dark conditions [5-9] but to learn about charge transport during solar cell operation, these devices must also be studied under illumination. Although studies of these devices have been reported [10], recombination rates based on these devices have not been reported. We note that the morphology in a FET configuration is different from that in a solar cell configuration, and that this could place limits on the interpretation of these results with reference to charge transport in solar cells. Nevertheless, a wealth of information is available from FET-based measurements.

To determine recombination rates and characterize other parameters of interest in this system, three basic measurements must be made: the output characteristic and the transfer characteristic (for both pFET and nFET modes), and a two-terminal measurement (an ungated source-drain I-V sweep). These parameters must be measured both in the dark and under illumination, preferably AM1.5g, to determine the necessary parameters. From the transfer characteristics, the dark electron and hole mobilities can be found using the known equation for drain current in the linear region of the FET. Using the slope of the diode characteristic curve, the conductivity can be obtained. By combining the dark mobility values and the dark conductivity, the dark carrier concentrations for electrons and holes can be obtained as shown in equation 1 where σ is the conductivity, n_e and n_h are the electron and hole concentrations, and μ_e and μ_h are the electron and hole mobilities.

$$\sigma = q\left(n_e \mu_e + n_h \mu_h\right)$$ (1)

The electron and hole mobilities under illumination can again be extracted from transfer curves and the conductivity under illumination can be extracted from the diode characteristic. Combining these parameters, we can extract the excess carrier concentration as the expression for conductivity becomes equation 2 under illumination, as we know that the number of photogenerated electrons and the number of photogenerated holes must be equal, $\Delta n_e = \Delta n_h = \Delta n$.

$$\sigma_{ph} = q\left(\Delta n\left(\mu_{e,ph} + \mu_{h,ph}\right) + n_e \mu_{e,ph} + n_h \mu_{h,ph}\right)$$ (2)

By combining expressions for the carrier velocity and recombination rate (equations 3 and 4), we find an expression for the recombination rate (equation 5) as a function of measured parameters, device geometry, and electrical testing parameters where U is the recombination rate, Δn is the excess carrier concentration, μ is the carrier mobility, V is the applied voltage, L is the transit length of the device, τ is the transit time, $\vec{v}$ is the carrier velocity, and $\vec{E}$ is the electric field..

$$\vec{v} = \mu\vec{E}$$ (3)

$$U = \frac{\Delta n}{\tau}$$ (4)

$$U = \frac{\Delta n\ \mu\ V}{L^2}$$ (5)

EXPERIMENT

Ambipolar OTFTs were prepared using a standard bottom-gate, bottom-contact geometry with a p-Si substrate and 2000 Å of thermally grown silicon dioxide. Source and drain electrodes were defined through photolithography with device channel lengths ranging from 3 μm to 100 μm. The W/L was 1000 for all devices with channel lengths of 50 μm and smaller, while the 100 μm channel length devices had a W/L of 500. Aluminum was chosen for its ability to effectively inject both electrons and holes in the BHJ [6], and was thermally evaporated to a final thickness of 500 Å. To form the BHJ absorber layer, *Plextronics Plexcore PV 1000*, which is a commercially available P3HT:PCBM solution, was spun-cast at 1000 rpm for 60 seconds and annealed on a hot plate at 140° C for 20 minutes in a nitrogen atmosphere.

Measurements were performed in a *Desert Cryogenics* cryogenic probe station at vacuum better than 10^{-3} Torr. All electrical measurements were performed using an *Agilent 4155C* semiconductor parameter analyzer. When necessary, sample illumination was provided by an *Oriel Model 66907 and 66912* arc lamp using a 150 W ozone-free xenon lamp. The output spectrum was modified using an AM1.5g spectrum filter and the illumination intensity was 81 mW/cm^2. Typical output characteristics and transfer characteristics were measured in the dark and under 81 mW/cm^2 AM1.5g illumination for the ambipolar OTFTs in both pFET and nFET modes. In addition, current vs. voltage characteristics were measured with the gate disconnected both with and without illumination.

DISCUSSION

Output characteristics are shown in Figure 1 from an ambipolar OTFT with a channel length of 10 μm and W/L=1000. Typical ambipolar OTFT characteristics are seen with a diode current existing for low gate bias. As gate bias increases, the familiar linear and saturation characteristics are observed. At increased source-drain bias, there is the onset of the diode current resulting from the opposite carrier type being injected for the high field at the contacts. Linear mobilities are calculated for $|V_{DS}| = 30$ resulting in a hole mobility of 3.9×10^{-6} cm^2/V·s and an electron mobility of 7.8×10^{-6} cm^2/V·s.

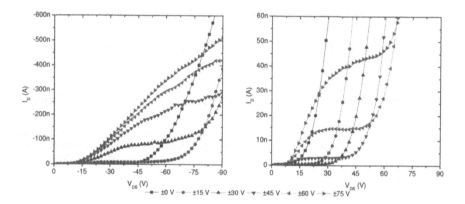

Figure 1. Output characteristics of an ambipolar OTFT. From left to right: pFET, nFET.

Output characteristics are shown in Figure 2 from the previous ambipolar OTFT under 81 mW/cm^2 AM1.5g illumination. Typical ambipolar OTFT characteristics can still be seen with an overall increase in drain current, indicating the ability to modulate the carrier concentration through an applied gate bias. Linear mobilities are calculated for $|V_{DS}| = 30$ resulting in a hole mobility of 3.8×10^{-6} cm^2/V·s and an electron mobility of 3.2×10^{-5} cm^2/V·s.

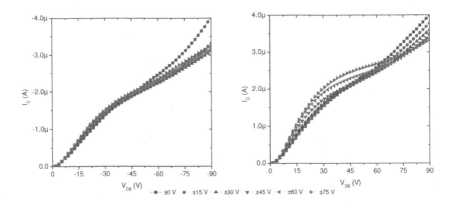

Figure 2. Output characteristics of an ambipolar OTFT under 81 mW/cm2 AM1.5g illumination. From left to right: pFET, nFET.

From the diode characteristics shown in Figure 3, dark and light conductivities have been calculated for $V_{DS} = 30$ to be 8.5×10^{-8} S/cm in the dark and 3.3×10^{-6} S/cm under illumination. This results in dark carrier concentrations of 1.1×10^{17} cm^{-3} for holes and of 1.5×10^{16} cm^{-3} for electrons. Under illumination, the excess hole and electron concentrations are equal at 2.3×10^{17} cm^{-3} each resulting in a recombination rate of 2.6×10^{19} cm^{-3} s^{-1}. Pivrikas et al. has reported a bimolecular recombination coefficient of 7×10^{-13} cm^3 s^{-1} which corresponds to a recombination rate of 5.8×10^{22} cm^{-3} s^{-1} [9]. These rates are not the total recombination rates but rather the recombination of separated electrons and holes being transported to the electrodes. We also note that the mobilities in organic semiconductors are carrier density dependent due to the nature of the energetic distribution of electronic states. However, this should not affect the values of recombination rate of separated carriers and the associated lifetime.

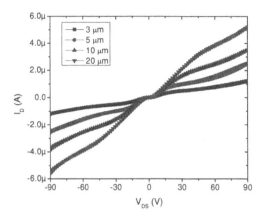

Figure 3. Left: Diode characteristics of an ambipolar OTFT in the dark and under 81 mW/cm2 AM1.5g illumination.

To further examine recombination effects in these BHJ films, diode characteristics were measured for ambipolar OTFTs of varying channel lengths. As most of the channel lengths have a constant W/L and, as a result, a different area for photon absorption, the photocurrent data was normalized to the absorption area. In organic semiconductors it is also known that the carrier mobility and thus conductivity and photocurrent are field dependant so the photocurrent measurements were normalized by dividing by the applied field. Figure 4 shows the photo current normalized to the absorption area and electric field for channel lengths ranging from 3 µm to 100 µm. For channel lengths ≤ 10 µm, the normalized photocurrent value is relatively unchanged at a value of about 1×10^{-8} A/V·cm. For channel lengths greater than 10 µm, the normalized photocurrent begins to decrease, indicating the onset of recombination. At low source-drain bias, contact effects are present so data was calculated at $V_{DS} = 30$. At these voltages, the effects of contact resistance are greatly reduced. Using this data and the recombination value found about for a 10 µm channel length device, the recombination time is estimated to be ≥ 8.8 ms.

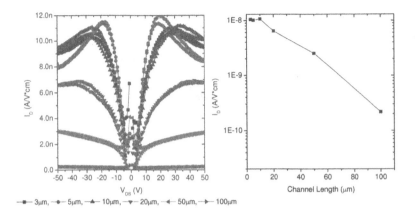

-■- 3µm, -●- 5µm, -▲- 10µm, -▼- 20µm, -◄- 50µm, -►- 100µm

Figure 4. Left: Diode characteristics of an ambipolar OTFT in the dark and under 81 mW/cm2 AM1.5g illumination normalized to absorption area and field. Right: Photocurrent normalized to absorption area and field for various channel lengths at $V_{DS} = 30$.

CONCLUSION

By measuring the output characteristic, transfer characteristic, and diode characteristic of ambipolar OTFTs under dark and illumination condition, we have determined some of the material parameters for this P3HT:PCBM material system that are useful in analyzing solar cells. Using these material parameters, an excess carrier concentration of 2.3×10^{17} cm^{-3} and a recombination rate of 2.6×10^{19} cm^{-3} s^{-1} have been calculated. Through studying the channel length dependence of the photocurrent, we have determined that significant recombination begins to occur at channel lengths greater than 10 µm and we have determined a recombination lifetime of ≥ 8.8 ms as well.

ACKNOWLEDGMENTS

The authors would like to thank members of the Dodabalapur Research Group at for useful discussions and the facilities staff at the Microelectronics Research Center for process related assistance. One of the authors would like to thank the Microelectronics and Computer Development Fellowship at The University of Texas at Austin for his financial support.

REFERENCES

1. M. A. Green, K. Emery, Y. Hishikawa and W. Warta, Progress in Photovoltaics: Research and Applications **18** (2), 144-150 (2010).
2. C. W. Tang, Applied Physics Letters **48** (2), 183-185 (1986).

3. G. Juška, K. Arlauskas, M. Viliūnas and J. Kočka, Physical Review Letters **84** (21), 4946 (2000).
4. R. Österbacka, A. Pivrikas, G. Juska, K. Genevicius, K. Arlauskas and H. Stubb, Current Applied Physics **4** (5), 534-538 (2004).
5. T. D. Anthopoulos, C. Tanase, S. Setayesh, E. J. Meijer, J. C. Hummelen, P. W. M. Blom and D. M. de Leeuw, Advanced Materials **16** (23-24), 2174-2179 (2004).
6. S. Cho, J. Yuen, J. Y. Kim, K. Lee and A. J. Heeger, Applied Physics Letters **89** (15), 153505-153503 (2006).
7. C. Rost, D. J. Gundlach, S. Karg and W. Riess, Journal of Applied Physics **95** (10), 5782-5787 (2004).
8. P. W. M. Blom, V. D. Mihailetchi, L. J. A. Koster and D. E. Markov, Advanced Materials **19** (12), 1551-1566 (2007).
9. A. Pivrikas, N. S. Sariciftci, G. Juscaronka and R. Österbacka, Progress in Photovoltaics: Research and Applications **15** (8), 677-696 (2007).
10. N. Marjanovic, T. B. Singh, G. Dennler, S. Günes, H. Neugebauer, N. S. Sariciftci, R. Schwödiauer and S. Bauer, Organic Electronics **7** (4), 188-194 (2006).

DSSC

Mater. Res. Soc. Symp. Proc. Vol. 1270 © 2010 Materials Research Society 1270-GG11-03

Role of Mesoporous TiO$_2$ Surface States and Metal Oxide Treatment on Charge Transport of Dye Sensitized Solar Cells

Mariyappan Shanmugam,[1] Braden Bills and Mahdi Farrokh Baroughi
Department of Electrical Engineering and Computer Science
South Dakota State University, Brookings, SD-57007, USA

ABSTRACT

Photovoltaic performance of dye sensitized solar cell (DSSC) was enhanced by 19 and 69 % compared to untreated DSSC by treating the nanoporous titanium dioxide (TiO$_2$) by ultra thin Aluminum oxide (Al$_2$O$_3$) and Hafnium oxide (HfO$_2$) grown by atomic layer deposition method. Activation energy of dark current, obtained from the temperature dependent current-voltage (I-V-T), of the untreated DSSC was 1.03 eV on the other hand the DSSCs with Al$_2$O$_3$ and HfO$_2$ surface treatment showed 1.27 and 1.31 eV respectively. A significant change in the activation energy of dark current, over 0.24 eV for Al$_2$O$_3$ treatment and 0.28 eV in case of HfO$_2$ treatment, suggest that density and activity of surface states on nanoporous TiO$_2$ was suppressed by ALD grown metal oxides to result improved photovoltaic performance. Further the enhanced DSSC performance was confirmed by external quantum efficiency measurement in the wavelength range of 350-750 nm.

INTRODUCTION

DSSCs have shown great potential to compete with conventional p-n junction solar cells in terms of conversion efficiency (η), cost, simple, fast and low-temperature fabrication procedures [1-3]. However, major challenges with DSSCs include narrow band absorption dyes, poor stability, lack of robustness in large scale production, and corrosive nature of the liquid electrolyte which affect the platinum coated electrode over time [4, 5]. Another significant problem associated with DSSCs is photogenerated carrier recombination loss that occurs at the interface between solid inorganic semiconductor, usually n type mesoporous TiO$_2$, organic dye, and liquid electrolyte [8, 9]. The J$_{SC}$, open circuit voltage (V$_{OC}$) and fill factor (FF) of DSSCs are greatly dependent on the density and activity of defects at the interface. Modeling and experimental evidences were presented to prove the charge transfer and recombination losses at the TiO$_2$/dye/electrolyte interface were the most dominant factor in DSSC performance [10].

A. Kay *et al.* has reported that the density of trap states and their activity can effectively be engineered by treating the mesoporous TiO$_2$ using wet chemical processed ZnO, TiO$_2$, ZrO$_2$, MgO, Al$_2$O$_3$, and Y$_2$O$_3$ [11]. E. Palomares *et al.* has reported significant enhancement in DSSC performance using wet chemical processed Al$_2$O$_3$ treated mesoporous TiO$_2$ shown by increased J$_{SC}$ from 8.1 to 10.9 mA/cm^2, increased V$_{OC}$ from 705 to 750 mV and improved efficiency from 3.8 to 5% [12]. Wet chemical processed TiCl$_4$ surface treatment on mesoporous TiO$_2$ also yielded a significant change in J$_{SC}$ from 9.5 to 11.2 mA/cm^2, V$_{OC}$ from 680 to 690 mV and improved cell efficiency from 4.3 % to 5.1 %, and was observed to change the morphology and surface area of the mesoporous TiO$_2$ resulting in enhanced dye loading capability of the photoelectrode [13]. Another

---

[1] Corresponding author: Tel.: +1(605)651-1804, Fax: +1(605)688-4401
E Mail: Mariyappan.Shanmugam@sdstate.edu (Mariyappan Shanmugam)

advanced technique to treat defective mesoporous TiO_2 is to use oxygen, nitrogen, hydrogen or argon gas phase plasma, to treat the TiO_2 sub-oxides such as Ti_2O_3 and TiO that are formed during high temperature sintering process resulting in slightly enhanced DSSC performance [14].

Recently, ultra thin (a few atomic layers) Al_2O_3 layers, grown by ALD technique, have been used to treat $TiO_2/dye/CuI$ and $TiO_2/dye/CuInS_2$ interfaces in solid state DSSCs and enhancements in conversion efficiency of over 20 % were observed [15]. ALD method is advantageous over wet chemical method for treating defective TiO_2 surfaces with metal oxides due to the higher purity of gas precursors, faster diffusion of gas precursors into mesoporous network, and better conformallity and controllability of metal oxide thickness. This paper describes the role of ultra thin Al_2O_3 and HfO_2 layers grown by ALD at the $TiO_2/dye/electrolyte$ interface on the performance of liquid electrolyte based DSSCs. Further, the first experimental demonstration of HfO_2 surface treatment on mesoporous TiO_2 to achieve high performance DSSCs is reported.

EXPERIMENTAL

The detailed reference DSSC fabrication procedure followed in this letter was reported by M. Shanmugam et. al. [16]. For fabrication of interface treated DSSCs, Al_2O_3 and HfO_2 ultra thin layers of 5, 10 and 20 cycles were deposited on mesoporous TiO_2 by ALD (Savannah 100, Cambridge NanoTech). Tri methyl aluminum and hafnium tetra chloride gases, as the Al and Hf precursors respectively, and water vapor (H_2O) as the oxygen precursor, were applied sequentially into the deposition chamber to grow Al_2O_3 and HfO_2 on mesoporous TiO_2 at 200 °C. This process was very short, where each cycle was 10 seconds long. The remaining fabrication procedures were the same as those for the reference DSSC. DSSCs fabricated using 5, 10 and 20 cycles of Al_2O_3 on mesoporous TiO_2 were designated device 1, 2 and 3 respectively; while, for HfO_2, 5, 10 and 20 cycles were designated as devices 4, 5 and 6 respectively.

Current-voltage (I-V) characteristics of the DSSCs were characterized by an Agilent 4155C semiconductor parameter analyzer. Illuminated I-V measurements were performed using an AM1.5 filtered Xenon arc lamp. NREL calibrated Hamamatsu S1133 photodetector, similar to a N719 based DSSC that absorbs photons in 350–800 nm wavelength window, was used to calibrate the illumination power at 100 mW/cm^2. The same photodetector was used as the reference cell to calibrate EQE measurements in the wavelength window of 350 – 800 nm. To understand the effect of ultra thin Al_2O_3 and HfO_2 on the performance DSSCs, I-V-T measurements were carried out for the reference DSSC and DSSCs with 20 cycles of Al_2O_3 and 5 cycles of HfO_2 in the temperature range of 303-328K and activation energies of the dark saturation current were obtained by fitting the exponential part of the curve, in the medium forward bias region, into a single diode model.

RESULTS AND DISCUSSION

Fig. 1 shows the illuminated I-V characteristics of the reference DSSC along with (a) Al_2O_3 and HfO_2 treated DSSCs. The illuminated I-V characteristics showed the carrier transport at the interface $TiO_2/dye/electrolyte$ was not blocked despite the very small electron affinities of Al_2O_3 and HfO_2 (1.35 and 2 eV respectively) compared to that of TiO_2 (3.9 eV). This might be explained by tunneling of electrons through the ultra thin metal oxide layers.

Devices 1, 2 and 3 had significantly higher J_{SC} (9 to 40%) with slightly reduced V_{OC} as compared to the reference DSSC as shown in Figure 2(a). Devices 4, 5, and 6 had significantly higher J_{SC} (35% to 67%), while V_{OC} increased for Devices 4 and 5 compared to the reference cell as shown in Figure 2(b). The measured cell parameters V_{OC}, J_{SC}, FF, η and efficiency enhancement with respect to the reference cell (E) of devices 1 through 6 are listed in Table 1. Devices 3 and 4 had the best illuminated I-V enhancement. No trend in FF values was observed. Further increase in thickness of the HfO_2 appeared to block carrier transport as can be seen in the suppressed photocurrent density.

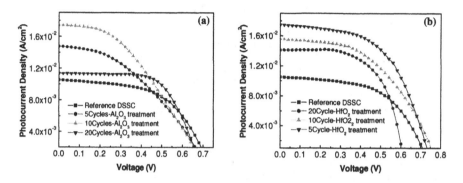

Fig. 1 Illuminated I-V characteristics of the reference and DSSCs with 5, 10 and 20 cycles of (a) Al_2O_3 treated mesoporous TiO_2 and (b) HfO_2 treated mesoporous TiO_2.

Table I: Illuminated solar cell parameters of reference, Al_2O_3 and HfO_2 treated DSSCs.

DCCSs	Type of Photoelectrode	J_{SC} (mA/cm^2)	V_{OC} (mV)	FF	η (%)	E (%)
Reference	TiO_2	10.5	730	0.55	4.2	--
Device 1	TiO_2 / Al_2O_3 (5 Cyc.)	14.7	708	0.42	4.2	0
Device 2	TiO_2 / Al_2O_3 (10 Cyc.)	17.4	700	0.40	4.9	17
Device 3	TiO_2/ Al_2O_3 (20 Cyc.)	11.4	684	0.64	5.0	19
Device 4	TiO_2 / HfO_2 (5 Cyc.)	17.5	745	0.55	7.1	69
Device 5	TiO_2 / HfO_2 (10 Cyc.)	15.6	765	0.49	5.8	38
Device 6	TiO_2/HfO_2 (20 Cyc.)	14.2	611	0.61	5.3	26

Fig. 2 shows the EQE of devices 3 and 4, which showed better performance than other DSSCs, along with that of the reference DSSC. The overall shape of all EQE curves was very similar. This suggests that the energy structure of N719 molecule in TiO_2/dye, TiO_2/Al_2O_3/dye, and TiO_2/HfO_2/dye systems was very similar and hence the absorption spectra in all three cases were similar. Further, HfO_2 treated DSSC showed a better EQE spectrum than that of the Al_2O_3 treated DSSC, consistent with the results from illuminated I-V characteristics.

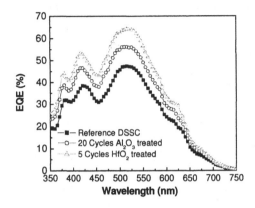

Fig. 2 EQE of the reference and the DSSCs with 20 cycles of Al_2O_3 and 5 cycles of HfO_2 treated mesoporous TiO_2

Since the dark I-V response of a DSSC originates from electron injection from the E_C of TiO_2 to I^-/I_3^- electrolyte through two serial processes: 1) electron trapping from E_C to surface states in mesoporous TiO_2 and 2) electron donation from TiO_2 surface states to I_3^- ions of electrolyte to form I^- ions (reduction process). Suppressing this injection, which is present in the forward bias region under both dark and illuminated conditions, can lead to higher J_{SC} and V_{OC} values in DSSCs. Incorporation of ultra thin metal oxide layers between TiO_2 and liquid electrolyte affects both the capture rate of electrons from E_C of the TiO_2 to surface states and the transfer rate of electrons from surface states to electrolyte.

Fig. 3 shows the I-V-T characteristics of the (a) reference DSSC, (b) DSSC with 20 cycles of Al_2O_3 interfacial layer (c) DSSC with 5 cycles of HfO_2. Activation energies of electron transport in the reference, DSSCs with 20 cycles of Al_2O_3 and 5 cycles of HfO_2 interface layers were 1.03, 1.27 and 1.31eV respectively as shown in the insets of Fig 3 (a), (b) and (c).

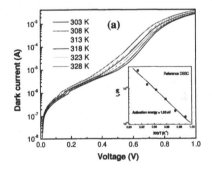
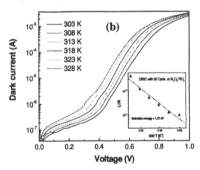

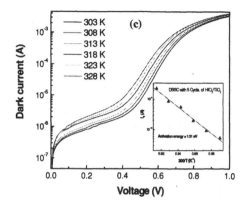

Fig. 3. I-V-T measurements and the activation energies (insets) of (a) reference DSSC (b) DSSC with 20 cycles of Al₂O₃ (c) DSSC with 5 cycles of HfO₂.

We believe that the activation energy of the dark current represents the energy difference between the conduction band of the TiO$_2$ and active surface states which participate in the reduction process (donate electrons to I$_3^-$ ions to form I$^-$ ions), where a large activation energy lower electron capture probability by active surface states and hence a lower recombination current. The results suggest that the incorporation of the ALD metal oxides at the TiO$_2$/dye/electrolyte interface increases the energy separation of TiO$_2$ conduction band than the active surface states and hence reduces the probability of electron injection from the TiO$_2$ conduction band to the liquid electrolyte through the surface states.

Dark and illuminated I-V results presented in this work accompanied by illuminated I-V results reported by Y. Noma *et al.* suggest that the interfacial metal oxide layers can suppress the transfer rate of electrons from TiO$_2$ to electrolyte and lead to better DSSC efficiencies [17]. The enhancement in performance of DSSCs by ultra thin metal oxide treatment at TiO$_2$/dye/electrolyte interface is in agreement with the results reported by A. Kay *et al.* and E. Palomares *et al.* who used wet chemical processed thin metal oxide interfacial layers to enhance electronic quality of TiO$_2$/dye/electrolyte interface , [12, 13]. Therefore it appears that the ALD method can be advantageous over wet chemical processing method for realizing thin interfacial metal oxides for DSSC applications due to faster processing and higher level of control over the thickness, conformallity, and purity of the layer.

CONCLUSIONS

In conclusion, ALD grown ultra thin Al$_2$O$_3$ and HfO$_2$ films at the TiO$_2$/dye/electrolyte interface in DSSCs resulted in suppressed dark saturation current, enhanced short circuit current density, and improved device performance. A significant increase in activation energy (over 0.2 eV) of the dark saturation current due to ALD oxide treatment confirms a major reduction in the recombination current of the surface treated DSSCs. Results suggested that ALD grown metal oxide layers can replace the wet chemical processed materials for mesoporous TiO_2 surface treatment and the quick process time would lead to high throughput of DSSC fabrication.

ACKNOWLEDGEMENT

Authors would like to thank NSF-EPSCoR-0554609 program and American Science and Technology for partial funding of this research, Nanofabrication Center of the University of Minnesota for providing the ALD facility and National Renewable Energy Laboratory for calibrating the reference photodetectors for illuminated I-V measurements.

REFERENCES

[1] B. O'Regan, M. Gratzel, Nature **353**, 737 (1991).

[2] M. Durr, A. Schmid, M. Obermaier, S. Rosselli, A. Yasuda and G. Nelles, Nature Materials **4**, 607 (2005).

[3] J. R.Durrant and S. A. Haque, Nature Materials **4**, 362 (2003).

[4] M. Gratzel, C. R. Chimie 9, 578 (2006).

[5] H. Nusbaumer, S. M. Zakeeruddin, J. E. Moser, and M. Gratzel, Chem. Eur. J. **9**, 3756 (2003)

[6] D. Cahen, G. Hodes, M. Gratzel, J. Franois Guillemoles, and I. Riess, J. Phys. Chem. B , **104**, 2053 (2000).

[7] S. Sakaguchi, H. Ueki, T. Kato, T. Kado, R. Shiratuchi, W. Takashima, K. Kaneto, S. Hayase, Journal of Photochemistry and Photobiology A: Chemistry **164**, 117 (2004).

[8] B. A. Gregg, F. Pichot, S. Ferrere, and C. L. Fields J. Phys. Chem. B 105, 1422 (2001).

[9] V. Yong, S.Tiong Ho, and R. P. H. Chang, Appl. Phys. Lett. **92**, 143506 (2008).

[10] M. Liberatore, L. Burtone, T. M. Brown, A. Reale, A. Di Carlo, F. Decker, S. Caramori, and C. A. Bignozzi, Appl. Phys. Lett. **94**, 173113 (2009).

[11] A.Kay and M. Gratzel, Chem. Mater., 14, 2930 (2002).

[12] 14E. Palomares, J. N. Clifford, S. A. Haque, T. Lutz and J. R. Durrant, Chem. Commun., 14, 1464 (2002).

[13] P. M. Sommeling, B. C. O'Regan, R. R. Haswell, H. J. P. Smit, N. J. Bakker, J. J. T. Smits, J. M. Kroon, and J. A. M. van Roosmalen, J. Phys. Chem. B **110**, 19191 (2006).

[14] Y. Kim, B. J. Yoo, R. Vittal, Y. Lee, N.G. Park, K.J. Kim, Journal of Power Sources **175**, 914 (2008).

[15] F.Lenzmann, M.Nanu, O.Kijatkina, A.Belaidi, Thin Solid Films **451–452**, 639 (2004).

[16] M. Shanmugam, M. Farrokh Baroughi and D. Galipeau, Electronics Letters **45, 648** (2009).

[17] Y. Noma, T. Kado, D. Ogata, Y. Hara and S. Hayase, Japanese Journal of Applied Physics **47**, 505(2008).

Morphology

Mater. Res. Soc. Symp. Proc. Vol. 1270 © 2010 Materials Research Society

Investigation of Local Dynamics on the Sub-micron Scale in Organic Blends using an Ultrafast Confocal Microscope

G. Grancini[1], D. Polli[1], J. Clark[1], T. Virgili[2], G. Cerullo[1], G.Lanzani[3]

[1]Dipartimento di Fisica, Politecnico di Milano, Piazza L. da Vinci 32, 20133 Milano, Italy

[2]IFN-CNR, Dipartimento di Fisica, Politecnico di Milano, Piazza L. da Vinci 32, 20133 Milano, Italy

[3]Center for Nano Science and Technology of IIT@POLIMI, via Pascoli 70/3 20133 Milano, Italy

ABSTRACT

We introduce a novel instrument combining femtosecond pump-probe spectroscopy and confocal microscopy for spatio-temporal imaging of excited-state dynamics of phase-separated polymer blends. Phenomena occurring at interfaces between different materials are crucial for optimizing the device performances, but are poorly understood due to the variety of possible electronic states and processes involved and to their complicated dynamics. Our instrument (with 200-fs temporal resolution and 300-nm spatial resolution) provides new insights into the properties of polymer blends, revealing spatially variable photo-relaxation paths and dynamics and highlighting a peculiar behaviour at the interface between the phase-separated domains.

INTRODUCTION

Plastic optoelectronics and photonics are based on blends of organic semiconductors, properly engineered in order to obtain samples with enhanced or new properties respect to the characteristics of their constituents. In such samples, phenomena occurring at interfaces between the two materials ultimately determine the desired device performances but are extremely complex and yet poorly understood. The photogenerated singlet excitons can undergo a variety of processes, such as mono- and bimolecular decay, internal conversion, energy and charge transfer or intersystem crossing. Ultrafast optical spectroscopy provides rich information on the photophysics of organic semiconductors [1], but in standard setups it has a limited spatial resolution (≈ 100 μm), so that the experimental results are averaged over many mesoscopic domains. Microscopy techniques, on the other hand, can provide very high spatial resolution, but they have no access to the excited state dynamics and to the relaxation paths which dictate the ultimate device performance. For these reasons, we have developed an instrument combining broadband femtosecond pump-probe spectroscopy with confocal microscopy, delivering simultaneously high temporal and spatial resolution. This tool provides new insight into the properties of polymer blends by directly measuring the sample dynamics at the interfaces between the different domains.

EXPERIMENT

The ultrafast confocal microscope setup is sketched in Figure 1. It is driven by 10-µJ, 150-fs pulses at 800 nm wavelength and 1-kHz repetition rate delivered by a commercial Ti:sapphire regeneratively-amplified laser. Pump pulses at 400 nm are produced by second harmonic generation, while probe pulses are created by filtering with 10-nm interference filters a white-light continuum generated in a sapphire plate. Pump and probe pulses, synchronized by a delay line, are collinearly recombined and focused on the sample by an air microscope objective with 100× magnification and 0.75 numerical aperture. The sample films are prepared on a reflective metal substrate, so that the reflection becomes equivalent to a double-pass transmission. The reflected probe is focused on the 50-µm core of an optical fiber, serving as the pinhole of the confocal microscope, and detected by a photomultiplier. By raster scanning the sample, mounted on a piezotranslator, we can acquire four-dimensional images of the differential transmission signal $\Delta T/T$ as a function of probe wavelength, pump-probe delay and x-y sample position. The instrument has a temporal resolution of $\approx$150 fs and a spatial resolution of $\approx$300 nm, with a sensitivity up to $\Delta T/T \approx 10^{-4}$.

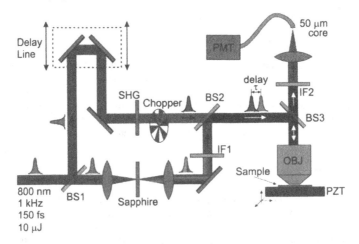

Figure 1. Experimental setup of the ultrafast confocal microscope. BS1, BS2, BS3: beam splitters; SHG: second-harmonic generation crystal; IF1, IF2: interference filters; OBJ: microscope objective; PZT: piezotranslator; PMT: photomultiplier.

RESULTS AND DISCUSSION

We have applied the ultrafast confocal microscope to map excited state dynamics in thin films of different polymer blends. The first is a thin film of poly(9,9-dioctylfluorene) (PFO), blended with polymethylmethacrylate (PMMA, 10% wt. PFO in PMMA), commonly used for amplification and all-optical switching in plastic optical fibers. PFO (see its chemical structure in Fig. 2(d)) is a blue-emitting polymer, with an absorption maximum at 385 nm, while PMMA is transparent at our pump wavelength. PFO can display different phases [2], namely aggregated and isolated; their $\Delta T/T$ spectra, measured with a standard pump-probe setup, are plotted as a reference in Figure 2(d). Note that in the 480-610 nm wavelength region the aggregated phase displays $\Delta T/T<0$ due to photo-induced absorption (PA) from the generated charges, while the isolated one shows $\Delta T/T>0$ due to stimulated emission SE from the singlet states.

Our experimental results are plotted in Figure 2. In panel (a) we show a 30x30 μm^2 $\Delta T/T$ image collected at 510 nm probe wavelength and 200 fs pump-probe delay. Figure 2(b) shows a 10x10 μm^2 zoom of the central-left part of Fig. 2(a). The film presents a phase separation [3] between the dark PMMA host matrix and the bright PFO disks of various shapes and sizes. Furthermore, the PFO islands display a spatially varying $\Delta T/T$ signal, with a negative area in the centre (blue in the figure), surrounded by a positive interfacial ring (red). The change of sign is indicative of the two different PFO phases: aggregated in the centre and isolated in the peripheral rim, as explained by Fig. 2(d). Standard pump-probe experiments would typically interrogate the whole scanned area thus averaging out these subtle differences. Our interpretation is confirmed by the image at $\lambda=640$ nm probe wavelength in Figure 2(c): here the aggregated PFO (in the centre) still displays a negative $\Delta T/T$ signal, while the isolated PFO (at the rim) has now a vanishing signal, in agreement with Fig. 2(d). Note that in Figure 2(c) the isolated PFO phase is much harder to recognize, highlighting the importance of our multi-dimensional probing. Note also that we could not detect any change in the linear (steady-state) optical properties of the

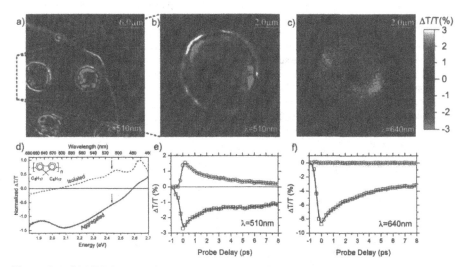

Figure 2. (a, b) $\Delta T/T$ images of a PFO/PMMA blend at 510 nm probe wavelength, showing PFO chain isolation ($\Delta T/T>0$, red) at the interface between the inner aggregated PFO ($\Delta T/T<0$, blue) and the outer inert PMMA matrix ($\Delta T/T\cong0$,dark). (b) Zoom of region (a). (c) Same as (b) but at 640 nm probe wavelength. (d) Normalized $\Delta T/T$ spectra of the isolated and aggregated PFO chains, showing the change of sign at the monitored 510 nm probe wavelength; inset: PFO chemical structure. (e, f) Pump-probe dynamics at 510 nm (e) and 640 nm (f) wavelengths in prototypical isolated (red) and aggregated (blue) areas.

border region of the PFO domains respect to the homogeneous inner part. The $\Delta T/T(\lambda)$ dynamics collected at two fixed spots on the sample in the centre and in the rim of the PFO domain, are shown in Figure 2(e) and (f) at the two different probe wavelengths, and are consistent with those previously measured on aggregated and isolated phases.

By deconvolution with the measured point-spread function, we can estimate the width of the isolated PFO rings to be variable from few tens of nanometers to 500nm, with an average value of $\approx$300 nm. The size of this interface region is controlled by a complex interplay of thermo-dynamical processes which regulate the PFO/PMMA miscibility and is in general a crucial factor in determining the performance of optoelectronic devices based on polymer blends. These results demonstrate that the different local environment of PFO and its peculiar inter-chain interactions influence the molecular photophysics on very small domains.

The second sample we studied is a drop-cast film obtained by blending a regioregular poly(3-hexylthiophene) (P3HT) polymer with 1-(3-methoxycarbonyl)propyl-1-phenyl-[6,6]C60 (PCBM). These blends are commonly termed bulk heterojunctions and consist of intermixed nanoscale domains of the two components, owing to the phase separation occurring between the two materials [4]. The blend morphology can be influenced by the preparation conditions (film thickness, thermal treatments, solvents…) and provides preferential percolation paths for electrons and holes in the fullerene and polymer domains, respectively. While the electronic structure of the pristine materials has been well characterized, the interfacial states in the blends are not fully understood, although they critically determine the charge transfer efficiency. Figure 3(a) shows a $\Delta T/T$ image of a P3HT/PCBM blend at 200-fs delay and 540 nm probe wavelength, at which the two components of the blend display different transient signals: P3HT shows $\Delta T/T>0$, due to photo-bleaching of the ground state absorption (dashed red line in Fig. 3(b)) [5]; PCBM shows $\Delta T/T<0$, due to PA from the excited state (solid blue line in Fig. 3(b)). Our instrument allowed us to map phase separation with high contrast, highlighting PCBM-rich circular areas (blue), with size of hundreds of nanometers immersed, in a P3HT-rich environment (red). At the interfaces between P3HT and PCBM, we observed a transient signal that is not simply the linear superposition of the individual species; this is due to the charge separation process, which is here mapped in real time and space.

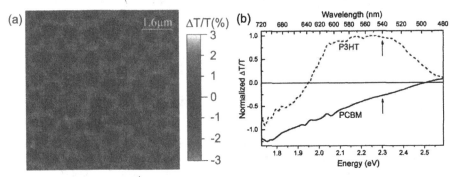

Figure 3. (a) $\Delta T/T$ images of a P3HT/PCBM blend at 540 nm probe wavelength, showing PCBM-rich circular areas ($\Delta T/T<0$, blue) immersed in P3HT-rich background ($\Delta T/T>0$, red). (b) Normalized $\Delta T/T$ spectra of the two species, explaining the change of sign at the monitored 540 nm probe wavelength.

CONCLUSIONS

In conclusion, we demonstrated a novel instrument enabling to map the excited-state dynamics in polymer blends with diffraction-limited resolution, showing their dependence on the sub-micron local morphology and composition. These measurements establish a direct link between the observed dynamical optical properties of the blends and the environment in which they occur, providing a completely new insight into their mesoscopic structure, which is of great importance for material scientists who strive to control the morphology and the supramolecular organization in order to optimize the device performance. We envisage that ultrafast confocal microscopy will become a standard tool for the characterisation of polymer blends.

REFERENCES

1. G. Lanzani, G. Cerullo, D. Polli, A. Gambetta, M. Zavelani-Rossi, C. Gadermaier, Phys. Stat. Sol. (a) 201, 1116 (2004).
2. T. Virgili, D. Marinotto, C. Manzoni, G. Cerullo and G. Lanzani, Phys. Rev. Lett. 94, 117402 (2005).
3. J. Chappell and D. G. Lidzey, J. Microsc. 209, 188–193 (2003).
4. H. Ohkita *et al.*, J.Am. Chem. Soc. 130, 3030 (2008).
5. O. J. Korovyanko, R. Oesterbacka, X. M. Jiang, Z. V. Vardeny, Phys. Rev. B 64, 235122 (2001).

Poster Session II

Mater. Res. Soc. Symp. Proc. Vol. 1270 © 2010 Materials Research Society   1270-HH14-07

Fabrication and Characterizations of Poly(3-hexylthiophene) Nanofibers

Surawut Chuangchote, Michiyasu Fujita, Takashi Sagawa[*], and Susumu Yoshikawa[*]
Institute of Advanced Energy, Kyoto University, Gokasho, Uji, Kyoto 611-0011, Japan
E-mail: t-sagawa@iae.kyoto-u.ac.jp; s-yoshi@iae.kyoto-u.ac.jp

ABSTRACT

Conductive polymer nanofibers with the average diameters in the range of 60 nm - 2 µm were fabricated by electrospinning of a mixture of poly(3-hexylthiophene) (P3HT) and polyvinylpyrrolidone (PVP) in a mixed solvent of chlorobenzene and methanol. Beaded fibers and/or uniform, smooth-surface fibers were successfully fabricated. The average diameter of the as-spun fibers decreased and the color of as-spun fibers changed with decreasing the concentration of P3HT or PVP. After the removal of PVP from as-spun fibers by Soxhlet extraction, pure P3HT fibers were obtained as a spindle-like with groove-like morphological appearance which may be widely applicable for some specific applications, such as photovoltaic cells, thin film transistors, and light emitting diodes.

INTRODUCTION

Conductive polymers have been shown to have optical, mechanical and electronic properties with easy processing. Therefore, a range of electronic and optoelectronic devices have been fabricated from conductive polymers, such as field-effect transistors, photovoltaic cells, and electroluminescent diodes. Recently, organic one-dimensional (1D) nanomaterials, including nanofibers, are growing broad interest because of their specific properties, e.g. large surface to volume ratio, improved mechanical properties, and flexibility in surface functionalities [1]. Conductive polymers have been fabricated into nanofiber form by various methods. Among them, electrospinning which utilizes electrostatic force to produce continuous ultrafine fibers with diameters ranging from microns down to a few nanometers has become one of the simple techniques for fabrication of conductive polymer nanofibers [2,3]. Because of the limitations of solvents to dissolve the polymers and suitable molecular weight of the polymers for electrospinning, electrospinning of neat conductive polymers resulted in non-uniform fibers with large amount of beads [3,4]. Therefore, various techniques have been developed to solve such those problems.

We have reported a simple technique for improvement in the uniformity of electrospun nanofibers of a conductive polymer, poly[2-methoxy-5-(2′-ethylhexyloxy)-1,4-phenylene-vinylene] (MEHPPV), by blending with poly(vinyl pyrrolidone) (PVP) (see chemical structure in Figure 1(a)), an easily spinnable and easily extractable polymer [4-6]. Ultrafine MEH-PPV fibers could be obtained from electrospinning and subsequent Soxhlet extraction. Obtained fibers were applied to organic photovoltaic cells [7]. In this contribution, this technique was applied for the fabrication of another conductive polymer, poly(3-hexylthiophene) (P3HT) (see chemical structure in Figure 1(b)), which has been received much attention recently due to its specific properties and various applications [8-10], especially in organic photovoltaic cells . Effects of solution properties and spinning conditions on the morphological appearance and fiber diameters were also systematically investigated.

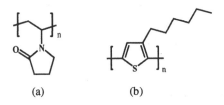

(a)   (b)

Figure 1. Chemical structures of (a) PVP and (b) P3HT

EXPERIMENT

P3HT was dissolved in a mixed solvent of chlorobenzene and methanol (85:15 v/v). PVP was added into the P3HT solution with vigorously stirring.

As-prepared solution was electrospun by using 60°C-heated syringe under an applied electrical potential of 15 kV over a fixed collection distance of 15 cm at room temperature. Collection time and feeding rate of the solution were fixed at 1 min and 1 mL h^{-1}, respectively. For removal of PVP from as-spun P3HT/PVP fibers, Soxhlet extraction was carried out with methanol at 75°C. The samples were then dried overnight at 60°C in vacuo. Morphological appearance of the as-spun fiber mats was observed by a scanning electron microscope (SEM), operating at an acceleration voltage of 10 kV. Ultraviolet-visible (UV-Vis) absorption and photoluminescence (PL) of the as-spun fibers were also investigated. The excited wavelength was 385 nm.

RESULTS AND DISCUSSION

Electrospinning of Pristine P3HT

SEM images of as-spun products from the electrospinning of pristine P3HT (10% w/v) in chlorobenzene under an applied electrical potential of 10 kV over a fixed collection distance of 15 cm are shown in Figure 2. The resulting products were beads and discrete rods with approximately 5 and 1 μm in diameters, respectively. This result indicates that the concentration of the solution was quite low, hence the low viscoelastic force that was not enough to prevent the partial breakup of the charged jet caused by the Coulombic repulsion during a flow instability of electrospinning [5]. Unfortunately, the concentration could not be increased much due to the solubility and the formation of gel-like solution which came from the high interchain interaction of P3HT.

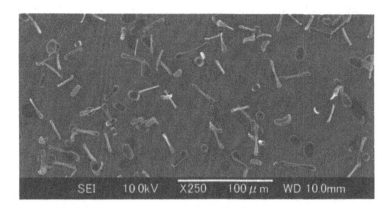

Figure 2. SEM image of the as-spun fibers from 10% w/v P3HT solution in chlorobenzene at an applied electrical potential of 10 kV. The collection distance was 15 cm. The scale bar is 100 μm.

Electrospinning of P3HT/PVP Blends

P3HT and PVP can be dissolved in different kinds of solvents. The solubility of P3HT in chlorobenzene and PVP in methanol is very well, while the opposite matching is insoluble. However, electrospun ultrafine fibers from PVP and also MEH-PPV/PVP solutions in the mixed solvent of chlorobenzene and methanol were found to be successfully fabricated [4-5]. Therefore, the mixture of chlorobenzene and methanol was chosen as the solvent for electrospinning of P3HT/PVP blended solutions. The volume ratio of chlorobenzene to methanol which could dissolve both of P3HT and PVP was in the range of 70:30 to 85:15.

Electrospinning of P3HT/PVP blend solutions (P3HT:PVP = 0.5:3 w/w, concentration of P3HT = 0.5% w/v) in the mixed solvents of chlorobenzene and methanol was carried out and the colors and diameters of obtained fibers were shown in Table 1. The diameter of as-spun fibers was found to increase while the number of beads was found to decrease with increasing composition of chlorobenzene because of increase in viscosity and decrease in dielectric constant. Regarding to color of fibers obtained, it was found that the color of electrospun fibers showed more uniform and darker with increasing the composition of chlorobenzene, because the solubility of P3HT in chlorobenzene is much higher than methanol. P3HT/PVP solution in chlorobenzene/methanol with a composition ratio of 85:15 v/v was chosen for following study because the solubility of P3HT in this solution was better than those of the other solutions and smooth fibers were obtained after electrospinning.

Table 1. Digital and SEM images of as-spun fibers from solutions of P3HT/PVP (P3HT:PVP = 0.5:3) in mixed solvents of chlorobenzene and methanol at various mixing ratio.

Image	Chlorobenzene/methanol			
	85:15	80:20	75:25	70:30
Digital				
SEM				

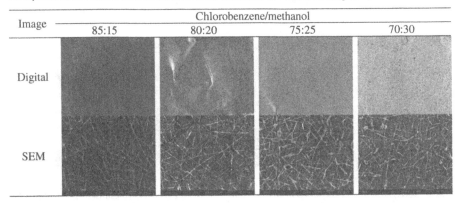

Beaded fibers and/or uniform, smooth-surface fibers were successfully fabricated by variation of solution conditions, i.e. concentration of P3HT and PVP (see examples in Figure 3). The average diameters of the fibers obtained could be varied in the range of nanometers to sub-micrometers (60 nm - 2 μm). It was found that the average diameter of the as-spun fibers decreased and the color of the as-spun fibers changed with decreasing the concentration of P3HT or PVP.

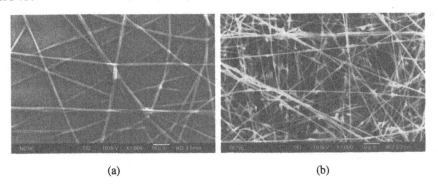

(a)   (b)

Figure 3. SEM images (scale bar = 10 μm) of as-spun P3HT/PVP fibers from solutions of (a) 5.1% w/v P3HT/PVP (P3HT:PVP = 0.1:5 w/w) and (b) 6% w/v P3HT/PVP (P3HT:PVP = 2:4 w/w) in chlorobenzene/methanol (85:15 v/v). The applied electrical potential was 15 kV over the fixed collection distance of 15 cm.

Extraction of PVP from P3HT/PVP Fibers

The removal of PVP from the as-spun P3HT/PVP fibers was carried out by using the Soxhlet extraction for 12 h. Figure 4 shows the typical SEM images of P3HT fibers after PVP had already been removed. The completely removal of PVP was confirmed by using UV-vis characterization.

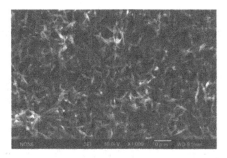

Figure 4. SEM image (scale bar = 10 μm) of P3HT fibers after Soxhlet extraction of P3HT/PVP composite fibers which were fabricated from solution of 6% w/v P3HT/PVP (P3HT:PVP = 2:4 w/w) (in Figure 3(b)).

After the removal of PVP, pure P3HT fibers showed a spindle-like with groove-like morphological appearance which has not been reported the open literature. Interestingly, there was an alignment of wrinkled surface in the spinning direction. The as-spun composite fibers showed the smooth surface, while after the removal of PVP, electrospun fibers changed to the aligned wrinkle surface. These results suggest that there was a phase separation of P3HT and PVP in the sub-micrometer scale occurred during instability and the confinement and electric field during electrospinning could enhance the orientation of this phase separation or also might be that of polymer chains in fiber direction. In addition, the removal of PVP also resulted the decreasing the diameter of fibers and thickness of fiber mats, and the enhancement of contacts parts of fiber mats.

CONCLUSIONS

Poly(3-hexylthiophene) (P3HT) and polyvinylpyrrolidone (PVP) composite nanofibers with the average diameters ranging in nanometers to sub-micrometers (60 nm - 2 μm) were fabricated by electrospinning of blended polymers in a mixed solvent of chlorobenzene and methanol. Beaded fibers and/or uniform, smooth-surface fibers were successfully fabricated. The average diameter of the as-spun fibers decreased and the color of as-spun fibers changed with decreasing the concentration of P3HT or PVP. After the removal of PVP from as-spun fibers by Soxhlet extraction, pure P3HT fibers were obtained as a spindle-like with groove-like morphological appearance which may be widely applicable for some specific applications, such as photovoltaic cells, thin film transistors, and light emitting diodes.

ACKNOWLEDGMENTS

This work was supported by grant-in-aids from Japan Society for the Promotion of Science (JSPS) under the JSPS Postdoctoral Fellowship for Foreign Researchers.

REFERENCES

1. S. Chuangchote, A. Sirivat, and P. Supaphol, *Nanotechnology* **18**, 145705 (2007).
2. S. Madhugiri, A. Dalton, J. Gutierrez, J. P. Ferraris, and K. J. Balkus, Jr., *J. Am. Chem. Soc.* **125**, 14531 (2003).
3. A. Babel, D. Li, Y. Xia, and S. A. Jenekhe, *Macromolecules* **38**, 4705 (2005).
4. S. Chuangchote, T. Sagawa, and S. Yoshikawa, *Jpn. J. Appl. Phys.* **47**, 787 (2008).
5. S. Chuangchote, T. Sagawa, and S. Yoshikawa, *Macromol. Symp.* **264**, 80 (2008).
6. S. Chuangchote, T. Sagawa, and S. Yoshikawa, *Mater. Res. Soc. Symp. Proc.* **1091E**, 1091-AA07-85 (2008).
7. S. Chuangchote, T. Sagawa, and S. Yoshikawa, *Mater. Res. Soc. Symp. Proc.* **1149E**, 1149-QQ11-04 (2008).
8. F. Padinger, R. S. Rittberger, and N. S. Sariciftci, *Adv. Funct. Mater.* **13**, 85 (2003).
9. X. Yang, J. Loos, S. C. Veenstra, W. J. H. Verhees, M. M. Wien, J. M. Kroon, M. A. J. Michels, and R. A. J. Janssen, *Nano Lett.* **5**, 579, (2005).
10. K. A. Singh, G. Sauve, R. Zhang, T. Kowalewski, R. D. McCullough, and L. M. Porter, *Appl. Phys. Lett.* **92**, 263303 (2008).

Mater. Res. Soc. Symp. Proc. Vol. 1270 © 2010 Materials Research Society   1270-HH14-18

Enhanced Performance from Acid Functionalized Multiwall Carbon Nanotubes in the Active Layer of Organic Bulk Heterojunction Solar Cells

Nasrul A. Nismy, A. A. Damitha T. Adikaari & S. Ravi P. Silva[*]
Nano-electronics centre
Advanced Technology Institute
University of Surrey, Guildford GU2 7XH, UK.

* Corresponding author email: s.silva@surrey.ac.uk

ABSTRACT

Solution-processable organic bulk-heterojunction photovoltaic devices have made great advances over the past decade. The concept, ultrafast photo induced electron transfer from a conjugated polymer to fullerene derivative molecules in bulk-heterojunction systems, leads to device efficiencies as high as 6%. Light absorption, charge separation and charge transport to electrodes are the most important steps in organic photovoltaic devices. The enhanced light absorption through thicker active layers results in more exciton creation, however, leads to increased recombination due to the relatively short exciton diffusion length. We fabricated poly(3-hexylthiophene)/ [6,6]-phenyl C_{61} butyric acid methyl ester bulk-heterojunction devices with multiwall carbon nanotubes in the active layer in a bid to address this deficiency. Functionalization of carbon nanotubes allows better dispersion in aromatic solvents, 1,2-dichlorobenzene in this study, and pristine multiwall nanotubes result in poorer dispersions. Organic photovoltaic devices fabricated with pristine multiwall carbon nanotubes in the active layer result in power conversion efficiencies ~1.4%, which show localized nanotube-rich areas in the active layer. Alternatively, acid functionalized nanotubes in the active layer results in efficiencies as high as 2.2 % with no distinct nanotube-rich sectors. The open circuit voltages of the devices show a dependency on the loading of nanotubes in the active layer. Further, the shunt resistances of the devices with carbon nanotubes decrease, which needs careful selection of the tubes depending on active layer thickness. This work compares the device performances in detail and identifies further improvements to conjugated polymer/fullerene derivative/multiwall carbon nanotubes hybrid photovoltaic systems.

INTRODUCTION

Photovoltaic devices based on organic materials offer great potential for easy device manufacturing on large area, light and flexible substrates at a low cost in comparison with silicon based inorganic solar cells [1-2].

The discovery of efficient charge transfer between the organic photoactive polymers and fullerene [3] modified the device architecture from bi-layer structures [4] to the bulk-heterojunction (BHJ) structures in organic photovoltaics (OPV) [5]. BHJ is a randomly phase separated, donor acceptor network. In BHJ based OPVs, the active layer consists of an intimate mixture of two different materials, one serving as an electron donor (D) and the other as an electron acceptor (A). Currently, the most efficient OPV devices are fabricated from conjugated organic materials and the most commonly used OPV material combination based on this concept is poly(3-hexylthiophene) (P3HT) and [6,6]-phenyl C_{61} butyric acid methyl fullerene (PCBM) which act as D and A layers respectively.

Carbon nanotubes have superior electron transport properties. In a carbon nanotube structure, every carbon atom is bonded to the nearest carbon atom leaving one electron in the P_z orbital. These electrons in the extended nanotube axis are delocalized and create an electron cloud. In this work, we have used multi wall carbon nanotubes (MWCNTs) to exploit the electronic properties of this delocalized electron cloud, with the conjugated organic molecules to enhance the charge transport properties in the OPV devices. Carbon nanotubes have been introduced in to the OPVs not only as a material in the active layer but also as the electron acceptor [6] and transparent electrodes [7]. Photovoltaic devices fabricated from MWCNTs incorporated active material composites have given a maximum reported efficiency of 2.0%[8].

Carbon nanotube dispersion is highly topical in contemporary nanotube literature. A well blended mixture of active layer is required for a uniform thin film and, it is an important factor in nano scale device fabrication procedure. Pristine MWCNTs are difficult to disperse in organic solvents, especially in aromatic solvents such as 1,2-dichlorobenzene (DCB) [9]. Therefore, in order to overcome some of the inherent difficulties associated with the MWCNTs, functionalization of nanotubes was considered. A major milestone in the nanotube chemistry was the development of an oxidation process for the nanotubes which involves extensive ultrasonic treatment in a mixture of strong acidic media and this process is called acid functionalization. Acid functionalized nanotubes according to this procedure retain their pristine electronic and mechanical properties [7, 10]. We report incorporation of acid functionalized MWCNTs (O-MWCNTs) into the OPV material combination, and the resulting device performance.

EXPERIMENTAL DETAILS

MWCNTs were purchased from Sigma Aldrich, U.K. Acid functionalization process was conducted by oxidizing the MWCNTs in concentrated acids. This acidic media comprised of 1:3 mixture of nitric and sulphuric acid. Then the solution was sonicated for 20 minutes and heated in an oil bath at 130 °C for two hours to eliminate impurities. The acid treated MWCNTs were thoroughly washed by deionized (DI) water. These nanotubes were centrifuged at 8500 rpm for 10 minutes, three times, to remove amorphous carbon and large aggregates. The centrifuged nanotubes were filtered by a polytetrafluroethylene membrane (0.2 μm) and thoroughly washed with DI water. Finally, the O-MWCNTs were solubilized in DI water. The concentration of this final solution was measured by evaporating a known volume of the solution. The typical diameter and length of the O-MWCNTs were 10 nm and few microns respectively, as published in a previous paper from the group [11]. The verification was via TEM analysis.

In order to prepare the O-MWCNTs incorporated active layer for OPV devices, initially, 1 mg of O- MWCNTs was dispersed in 1 ml of DCB and sonicated for one hour. Then, 15 mg of P3HT (Rieke metals, U.S.A.) was added to the solution and stirred for three hours before adding a further 15 mg of PCBM (Solenne, the Netherlands). The solution which contained

3.22% of O- MWCNTs by weight was allowed to stir overnight. Similarly, two other solutions were made to yield 1.64% and 0.33% of O- MWCNTs by weight. Also, pristine MWCNTs were used to make a P3HT:PCBM:MWCNT solution at 0.33% MWCNT loading.

Photovoltaic devices were fabricated on indium tin oxide (ITO) coated glass obtained from Luminescence Technologies Limited, Taiwan. The ITO coated substrates were cleaned through the ultrasonication process with acetone and methanol. These substrates were further treated with an oxygen plasma prior to the active layer deposition. The active layer composites, blends of P3HT: MWCNT: PCBM were deposited on to the ITO substrates by spin coating. A two step spin coating process was engaged. The first step was at 1000 rpm for 40 seconds and the second step was at 1500 rpm for 5 seconds. The spin coated devices were allowed to slow dry in closed Petri dishes at room temperature. Then, these devices were annealed at 125 °C for 10 minutes. The deposition of the active layer and annealing steps were performed in a nitrogen filled MBRAUN glove box. The exciton blocking layer (Bathocuproine (BCP), 5 nm) and the electrode (Al, 70 nm) were thermally evaporated using a shadow mask under vacuum at rates of 2 Å/s and 3 Å/s respectively. In the case of the devices without BCP, only the electrode, Al was deposited. The device structure, ITO/P3HT: O-MWCNTs: PCBM/BCP/Al comprised only of a hole blocking layer without an electron blocking layer, and has an active area close to 76 mm^2. Immediately after the device fabrication, current- voltage (I-V) measurements were performed at room temperature using a Keithley 2425 kit under a 300W Xe arc lamp Oriel simulator, fitted with an AM 1.5G filter, calibrated to an intensity of 1000Wm^{-2} . Transmission spectra of the active layer coated ITO substrates were taken using a Varian Cary 5000 spectrometer.

DISCUSSION

Photovoltaic devices with pristine MWCNTs and O-MWCNTs

The current density-voltage (J-V) curves for the devices made with 0.33 wt% of pristine MWCNTs (circles), and with O-MWCNTs for 1.64 wt% (squares) and 0.33 wt% (triangles) is shown in figure 1. The device fabricated with pristine MWCNTs shows an open circuit voltage (V_{oc}) of 0.54 V, short circuit current density (J_{sc}) of 5.80 mAcm^{-2}.The resulting power conversion efficiency (PCE) was 1.38% with fill factor (FF) of 44%. It is evident from the graphs that upon functionalization, the device performance have been improved. Better device parameters can be obtained even with higher concentrations of O-MWCNTs compared to the weight percentage of MWCNTs. This implies the significance of the functionalization of the tubes and the role of the treated nanotubes. 1.64 wt% of O-MWCNTs incorporated OPV devices resulted in V_{oc} of 0.53 V with an increase in J_{sc} to 6.05 mAcm^{-2}. The improved FF value of 51% is attributed to a low series resistance and high shunt resistance which lead to the PCE of1.65%. The progress achieved with the initial step of the treated nanotubes led to the next experimental section for the exact comparison of MWCNTs incorporated devices with O-MWCNTs.

The graph plotted in triangles in figure 1 is the J-V curve for the 0.33wt% of O-MWCNTs incorporated OPV device. It is the best device characteristics achieved with the presence of af-MWCNTs in the active layer found in literature. The considerable increase in

PCE of 2.28% was a combination of improvements in V_{oc} of 0.60 V, J_{sc} of 6.1 mAcm^{-2} and FF of 62% for the 0.33 wt% of af-MWCNTs devices.

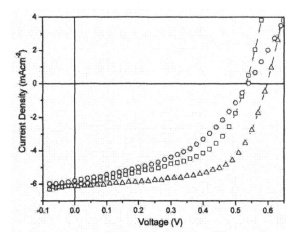

Figure 1. J-V characteristics of the device with 0.33 wt% of pristine MWCNT (circles), 1.64 wt% (squares) and 0.33 wt% (triangles) of O-MWCNTs respectively.

We claim that the significant enhancement in all the fundamental OPV parameters is due to the acid functionalization of MWCNTs. During the acid functionalization process, strong acids oxidize the nanotube surface and the open ends of the nanotubes are terminated by carboxylic acid (COOH) groups [12-15]. The presence of carboxylic acid groups leads to a reduction of van der Waals interactions between the nanotubes, which strongly facilitates the further separation of nanotube bundles into individual tubes. Additionally, the attached side groups render the tubes dispersible in aqueous or organic solvents [10].

Therefore, these dispersible nanotubes obtained from the functionalization process are the key factor for high performance in our devices. Increased J_{sc} suggests more exciton dissociation/extraction has taken place compared to the pristine MWCNTs incorporated devices and high charge transport may have been favorable for these dissociated excitons due to the presence of separated nanotubes in the active area. Also, this process is credited for a better film morphology required for a thin film device fabrication process. However, high nanotube concentrations in the active layer posed problems, and the devices made from 3.22 wt% of O-MWCNTs resulted in short circuits, destroying the diodes.

In order to investigate the difference in the J-V curves of the MWCNTs incorporated devices, optical transmission through the ITO/P3HT:MWCNTs: PCBM structures was obtained and significant differences not observed. Therefore, the similar absorption of light from pristine and O-MWCNTs introduced devices propose the acid functionalization process has not disrupted optical absorption process within the MWCNTs.

Furthermore, the hole blocking interface layer (BCP), employed was crucial for enhanced performance of these devices. Devices were fabricated without BCP for comparison and figure 2 shows the J-V characteristics for O- MWCNTs incorporated devices at 0.33 wt% with and without BCP. The V_{oc}, FF and PCE values decreased to 0.47 V, 53% and 1.48% respectively, while maintaining the J_{sc}, at 5.91mAcm^{-2}. The use of BCP with the active layer material combination of P3HT:O-MWCNTs:PCBM further increases the shunt resistance and is able to provide a smoother surface at the active layer/electrode contact. This helps minimize the surface roughness created by possible protruding O-MWCNTs and hence achieving a set of increased FF values.

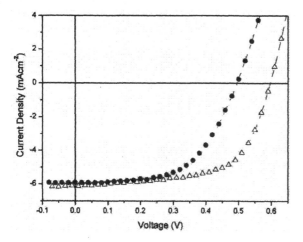

Figure 2. J-V characteristics of the devices with (triangles) and without (solid circles) bathocuproine (BCP) for the devices made with 0.33 wt% of O-MWCNT respectively.

CONCLUSIONS

A considerable improvement in device performance of organic hybrid solar cells was observed through the incorporation of acid functionalization of carbon nanotubes within the active layers. Enhanced performance from the acid treated nanotube incorporated devices compared to the pristine nanotube devices confirmed that relevant chemical modifications to the nanotube surface has the potential to transform with photoactive materials and contribute to the enhancement of fundamental device characteristics. Furthermore, PCE as high as 2.28% were obtained with the device structure of ITO/P3HT: O-MWCNTs: PCBM/BCP/Al, with the introduction of the exciton blocking layer, BCP.

ACKNOWLEDGEMENTS

This work is based on a project funded by E.ON AG, as part of the E.ON International Research Initiative. Responsibility for the content of this publication lies with the authors.

REFERENCES

[1]   C. J. Brabec, *et al.*, "Plastic Solar Cells," *Advanced Functional Materials,* vol. 11, pp. 15-26, 2001.

[2]   A. A. D. T. Adikaari, *et al.*, "Organic-Inorgnic Solar Cells: Recent Developments and Outlook," *IEEE,* J. Sel. Top Quantum Electron (to be published), DOI:10.1109/JSTQE.2010.2040464.

[3]   N. S. Sariciftci, *et al.*, "Photoinduced Electron Transfer from a Conducting Polymer to Buckminsterfullerene," *Science,* vol. 258, pp. 1474-1476, November 27, 1992 1992.

[4]   C. W. Tang, "Two-layer organic photovoltaic cell," *Applied Physics Letters,* vol. 48, pp. 183-185, 1986.

[5]   G. Yu, *et al.*, "Polymer Photovoltaic Cells: Enhanced Efficiencies via a Network of Internal Donor-Acceptor Heterojunctions," *Science,* vol. 270, pp. 1789-1791, December 15, 1995 1995.

[6]   E. Kymakis and G. A. J. Amaratunga, "Single-wall carbon nanotube/conjugated polymer photovoltaic devices," *Applied Physics Letters,* vol. 80, pp. 112-114, 2002; R. A. Hatton *et al.*, "Carbon nanotube: a multi functional material for organic optoelectronics," J. Mater Chem. vol. 18, pp. 1183-1192.

[7]   R. A. Hatton, *et al.*, "Oxidised carbon nanotubes as solution processable, high work function hole-extraction layers for organic solar cells," *Organic Electronics,* vol. 10, pp. 388-395, 2009.

[8]   S. Berson, *et al.*, "Elaboration of P3HT/CNT/PCBM Composites for Organic Photovoltaic Cells," *Advanced Functional Materials,* vol. 17, pp. 3363-3370, 2007.

[9]   E. Kymakis, *et al.*, "High open-circuit voltage photovoltaic devices from carbon-nanotube-polymer composites," *Journal of Applied Physics,* vol. 93, pp. 1764-1768, 2003; N. A. Nismy *et al.*, "Functionlaized Multiwall carbon nanotubes incorporated polymer/fullerene hybrid photovoltaics," Applied Physics Letters, vol. 97, pp. 033105-3,2010.

[10]   J. Zhang, *et al.*, "Effect of Chemical Oxidation on the Structure of Single-Walled Carbon Nanotubes," *The Journal of Physical Chemistry B,* vol. 107, pp. 3712-3718, 2003.

[11]   A. J. Miller, *et al.*, "Water-soluble multiwall-carbon-nanotube-polythiophene composite for bilayer photovoltaics," *Applied Physics Letters,* vol. 89, pp. 123115-3, 2006.

[12]   J. Chen, *et al.*, "Solution Properties of Single-Walled Carbon Nanotubes," *Science,* vol. 282, pp. 95-98, October 2, 1998 1998.

[13]   J. Liu, *et al.*, "Fullerene Pipes," *Science,* vol. 280, pp. 1253-1256, May 22, 1998 1998.

[14]   J. Shen, *et al.*, "Study on amino-functionalized multiwalled carbon nanotubes," *Materials Science and Engineering: A,* vol. 464, pp. 151-156, 2007.

[15]   B. I. Kharisov, *et al.*, "Recent Advances on the Soluble Carbon Nanotubes," *Industrial & Engineering Chemistry Research,* vol. 48, pp. 572-590, 2008.

Mater. Res. Soc. Symp. Proc. Vol. 1270 © 2010 Materials Research Society   1270-HH14-23

Effect of transparent electrode on the performance of bulk heterojunction solar cells

A. A. Damitha T. Adikaari[1*], Joe Briscoe[2], Steve Dunn[3], J. David Carey[1] and S. Ravi P. Silva[1]

[1] Nanoelectronics Centre, Advanced Technology Institute, University of Surrey, Guildford GU2 7XH, UK.

[2] Nanotechnology Centre, Cranfield University, Cranfield, MK43 0AL, UK.

[3] Materials Department, School of Engineering and Materials, Queen Mary, University of London, E1 4NS, UK.

*Corresponding author email:d.adikaari@surrey.ac.uk

ABSTRACT

We present a performance comparison of polythiophene/fullerene derivative bulk heterojunction solar cells fabricated on fluorinated tin oxide (FTO) and indium tin oxide (ITO) in the presence and absence of the commonly used poly(3,4-ethylenedioxythiophene): poly(styrenesulfonate) (PEDOT:PSS) hole extraction layer. From a potential commercial perspective the performance of cheaper and more readily available FTO compares well with the more expensive ITO in terms of measured device efficiency (FTO:2.8 % and ITO:3.1%). The devices show similar fill factors (FTO:63% and ITO:64%) with the same open circuit voltage of 0.6 V. The short circuit current density is lower for FTO devices at 7.5 mA/cm^2 which compares with 8.0 mA/cm^2 for ITO; a behaviour that is mainly attributed to the reduced optical transmission of the FTO layer. Importantly, these devices were part fabricated and wholly characterized under atmospheric conditions. The quoted device performance is the best reported for FTO based bulk heterojunction systems in the absence of the highly acidic PEDOT:PSS hole extraction layer, which is believed to degrade conductive oxides.

INTRODUCTION

Renewable energy is one of the key routes to minimize the adverse effects of carbon emission from energy conversion. The conversion of the sun's radiation to electricity by photovoltaics (PVs) offers one of the most credible alternatives for renewable energy needs provided the efficiency of the PVs can be increased while decreasing the cost. Silicon based PVs dominate 90% of the commercial market,[1] however the technology remains too expensive for mass uptake. Organic material-based PVs are cheaper and not as-energy intensive in terms of the production process. Furthermore, the possibility of facile device fabrication from solution makes the technology very attractive from a potential commercial perspective. However, the efficiencies and lifetimes of organic PV devices are inferior to their conventional inorganic semiconductor counterparts and are too low for commercial viability at present.

Interpenetrating charge donor-acceptor networks formed from a phase-segregated mixture of two semiconducting organic materials are considered to be the best architecture for fabrication of organic PVs to date[2, 3]. These interpenetrating networks are now known as bulk-heterojunction active layers. Bulk heterojunction PV devices fabricated have shown steady progress over the years [4-6], in terms of device efficiency, reaching a maximum of 7.73% in 2009 for a single junction device [7]. A vast majority of the bulk-heterojunction organic PV research has been on Poly (3-Hexylthiophene) (P3HT) donor and [6,6]-phenyl-C61-butyric acid methyl ester (PCBM) acceptor systems, coated on indium doped tin oxide (ITO) covered transparent substrates.

However, ITO is expensive to produce and it has been the standard to coat it with poly(3,4-ethylenedioxythiophene):poly(styrenesulfonate) (PEDOT:PSS) as a hole extraction layer for organic PV applications. Relatively few studies have focused on the development of alternative materials for use in OPV devices[8-12]. From these studies, fluorine-doped SnO_2 (FTO) is a possible alternative to ITO because SnO_2 films are inexpensive to produce and chemically and thermally stable. We report a performance comparison of ITO and FTO based P3HT and [6,6]-phenyl-C71-butyric acid methyl ester ($PC_{70}BM$) bulk heterojunction PVs with and without PEDOT:PSS with comparatively higher efficiencies than reported values.

EXPERIMENTAL

Photovoltaic devices were fabricated on ITO coated glass substrates (Luminescence Technology Corporation, Taiwan, 150 nm, 15Ω/□) and FTO coated glass substrates (Pilkington USA, TEC 15, 300 nm, 15Ω/□). The substrates were ultrasonically cleaned with acetone followed by methanol and treated with oxygen plasma for five minutes, before spinning 30 nm of PEDOT:PSS (H.C. Starck Clevios GmbH) at 3000 rpm for 60 seconds and drying at 140°C for 10 minutes. A reference set of samples was also used without the PEDOT:PSS layer. The active layers were spin coated on all four types of substrates from a blend of P3HT (24 mg):$PC_{70}BM$ (24mg) in 1 ml of 1-2 dichlorobenzene, stirred for a day on a magnetic stirrer with a stirring flea. A two step spin coating recipe was used yielding ~130 nm thick coating; the first at 800 rpm for 40 seconds and second at 1500 rpm for three seconds. The spin coated films were allowed to dry slowly in a closed glass Petri dish pair (60 mm diameter, 15 mm height) at room temperature and annealed at 125°C for 10 minutes. The active layer coating process was carried out in a nitrogen filled MBRAUN glove box with oxygen and water content less than 10 ppm. Bathocuproine (BCP, 8 nm) was used as a hole blocking layer and an Al (80 nm) electrode was thermally evaporated using a shadow mask at 8×10^{-6} mbar at the rate of 2 Å/s and 3 Å/s, respectively. The current voltage characteristics of the devices were measured using an Oriel 81160 solar simulator for AM 1.5G simulation while current-voltage measurements were collected with a Keithley 2425 source meter. A separate set of partial devices were characterised for optical transmission using a Varian Cary 5000 spectrophotometer.

DISCUSSION

Figure 1 shows the current density-voltage characteristics measured under dark condition for all for configurations of P3HT:$PC_{70}BM$ bulk heterojunction solar cells investigated. The turn-on voltages for the diodes are similar; however, the series resistances of devices on FTO are higher compared with the ITO based devices. Furthermore, the series resistances for both FTO and ITO based structures have increased in the presence of PEDOT:PSS. Although the less conductive PEDOT:PSS layer is thin (30 nm), the increase in series resistance for FTO based devices appears to be considerable.

68

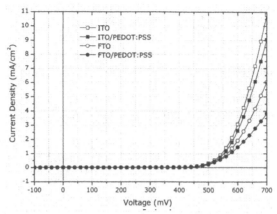

Figure 1. Current density-voltage characteristics in the dark for P3HT:PC$_{70}$BM bulk heterojunction solar cells.

Figure 2 shows the current density-voltage characteristics under AM 1.5G simulated irradiation for the same sets of devices. It can be observed that, contrary to many reports available, our fabrication procedure yields better output characteristics without the hole extracting PEDOT:PSS layer for both FTO and ITO systems. The short circuit current density (J_{sc}), open circuit voltage (V_{oc}), current density and the voltage at the maximum power point were extracted from the data, and were used to calculate the fill factor (FF) and the power conversion efficiency (η) of devices under test, parameters which are fundamental for comparison. Table 1 shows the comparison of performance parameters for the four types of devices tested in our study.

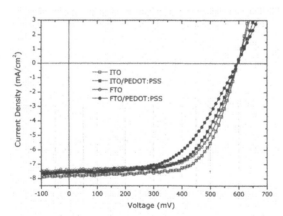

Figure 2. Current density-voltage characteristics under AM 1.5G simulated irradiation for P3HT:PC$_{70}$BM bulk heterojunction solar cells.

Table 1. Performance parameters of bulk heterojunction solar cells extracted from current-voltage characteristics under AM 1.5G simulated irradiation.

Architecture	Area (cm^2)	V_{oc} (V)	J_{sc} (mA/cm^2)	FF (%)	η (%)
Glass/ITO/P3HT:PCBM/BCP/Al	0.78	0.60	7.95	64	3.06
Glass/ITO/PEDOT:PSS/P3HT:PCBM/BCP/Al	0.78	0.60	7.64	60	2.73
Glass/FTO/P3HT:PCBM/BCP/Al	0.70	0.60	7.45	63	2.82
Glass/FTO/PEDOT:PSS/P3HT:PCBM/BCP/Al	0.70	0.60	7.59	54	2.45

The V_{oc}, which is fundamentally governed by the highest occupied molecular orbital of the acceptor and the lowest unoccupied molecular orbital of the donor, is the identical for all four devices. It can be seen that the best device is glass/ITO/P3HT:PCBM/BCP/Al, which does not utilize a hole extraction layer. Although the presence of PEDOT:PSS lowers the J_{sc} for ITO coated sample which is contrary to what is observed for devices using FTO. The change in J_{sc} is small with and without the hole extraction layer, and lower for FTO devices. This can be explained with the optical transmission spectra for FTO/ITO with and without active layer as shown in figure 3. The absorption of FTO is 2-5% higher in the region concerned (0.4 to 0.7 μm) compared with ITO, leading to lower J_{sc}. The higher absorption is partly due to twice the thickness of FTO compared to ITO, necessary for achieving similar sheet resistivities.

The FFs are better for cells without PEDOT:PSS, which suggests that the interfaces at ITO/FTO and BCP with active layer are optimal in this configuration. For both FTO and ITO, the FFs are similar suggesting that the work function difference between FTO (4.4 eV) and ITO (4.7 eV) [13] has an insignificant effect on the device architecture used. The quoted device performance is the best reported realistic efficiency for FTO based bulk heterojunction systems in the absence of the highly acidic PEDOT:PSS hole extraction layer which is believed to degrade conductive oxides. [12]

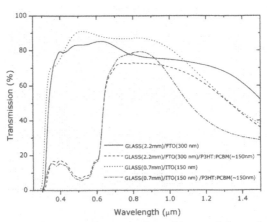

Figure 3. Optical transmission spectra for conductive oxide coated glass substrates and active layer coated substrates before BCP/Al deposition

CONCLUSIONS

A performance comparison of ITO and FTO based P3HT and [6,6]-phenyl-C71-butyric acid methyl ester ($PC_{70}BM$) bulk heterojunction PVs with and without PEDOT:PSS hole extraction layers was presented. It is concluded that FTO performs as well as ITO, with slightly lower J_{sc} owing to higher optical absorption of FTO leading to a reduction in available photons for photo-generation. Devices without the hole extraction layer yield better PVs under the device architecture used, with similar FFs for FTO as well as ITO, suggesting the work function differences of the two conducting oxides affect insignificantly for the device performance. The performance of cheaper and more readily available FTO compares well with the more expensive ITO in terms of measured device efficiency, which bodes well for future mass production.

ACKNOWLEDGMENTS

Part of this work is based on a project funded by E.ON AG, as part of the E.ON International Research Initiative. Responsibility for the content of this publication lies with the authors.

REFERENCES

[1]   R. F. Service, *Science* **319**, 718-720 (2008).
[2]   J. J. M. Halls, C. A. Walsh, N. C. Greenham, E. A. Marseglia, R. H. Friend, S. C. Moratti, and A. B. Holmes, *Nature* **376**, 498-500 (1995).
[3]   G. Yu, J. Gao, J. C. Hummelen, F. Wudl, and A. J. Heeger, *Science,* **270**, 1789-91 (1995).
[4]   G.Li, V. Shrotriya, H. Jinsong, Y. Yan, T. Moriarty, K. Emery, and Y. Yang, *Nature Materials,* **4**, 864-8, (2005).
[5]   K. Jin Young, L. Kwanghee, N. E. Coates, D. Moses, N. Thuc-Quyen, M. Dante, and A. J. Heeger, *Science,* **317**, 222-5 (2007).
[6]   G. Dennler, M. C. Scharber, and C. J. Brabec, *Advanced Materials,* **21**, 1323-1338 (2009).
[7]   H.-Y. Chen, J. Hou, S. Zhang, Y. Liang, G. Yang, Y. Yang, L. Yu, Y. Wu, and G. Li, *Nat Photon.* **3**, 649-653 (2009).
[8]   R. Valaski, R. Lessmann, L. S. Roman, I. A. Hümmelgen, R. M. Q. Mello, and L. Micaroni, *Electrochemistry Communications,* **6**, 357-360 (2004).
[9]   R. Valaski, F. Muchenski, R. Mello, L. Micaroni, L. Roman, and I. Hümmelgen, *Journal of Solid State Electrochemistry,* **10**, 24-27 (2006).
[10]   M. W. Rowell, M. A. Topinka, M. D. McGehee, H.-J. Prall, G. Dennler, N. S. Sariciftci, L. Hu, and G. Gruner, *Applied Physics Letters,* **88**, 233506 (2006).
[11]   F. Yang and S. R. Forrest, *Advanced Materials,* **18**, 2018-2022 (2006).
[12]   H. Kim, G. P. Kushto, R. C. Y. Auyeung, and A. Pique, *Applied Physics A: Materials Science and Processing,* **93**, 521-526 (2008).
[13]   A. Andersson, N. Johansson, P. Broems, N. Yu, D. Lupo, and W. R. Salaneck, *Advanced Materials,* **10**, 859-863 (1998).

Mater. Res. Soc. Symp. Proc. Vol. 1270 © 2010 Materials Research Society 1270-HH14-33

Interpretation of the C1s XPS Signal in Copper Phthalocyanine for Organic Photovoltaic Device Applications.

K. Nauka, Hou T. Ng and E.G. Hanson

Hewlett-Packard Laboratories, Hewlett-Packard Company, 1501 Page Mill Road, Palo Alto, CA 94304, USA.

ABSTRACT

Carbon 1s signal in the photoelectron spectrum of copper phthalocyanine has been resolved into four components representing two principal carbon positions within the copper phthalocyanine macrocycle and their shake-up transitions. In addition, the contribution of organic impurities frequently found in commercial CuPc materials has been accounted for with a high degree of accuracy by adding an extra component of the C1s signal. Its magnitude has been correlated with the intensity of aliphatic C-H vibrations observed in the IR spectrum demonstrating that IR absorption measurement can be successfully used for a routine evaluation of impurities in commercial CuPc materials used to fabricate organic photovoltaic devices.

INTRODUCTION

Copper phthalocyanine (CuPc) molecules form planar structures consisting of porphyrin-like rings surrounded by four peripheral benzene rings and central Cu atom. Because of the π-conjugation of the center carbon atoms CuPc exhibits semiconductor behavior that has found a number of potential applications in organic electronics [1,2]. In addition, CuPc is used as a major cyan pigment in a large number of applications ranging from printing on a paper to fabric dyeing [3]. Electronic applications require high purity CuPc materials with occasional stringently controlled doping providing the desired electrical conductivity. Mass-produced commercial CuPc may contain up to several percent of mostly organic impurities in form of undesirable manufacturing residues or intentional additives used to control handling and storage of the CuPc nanocrystalline powders. Although photoelectron spectroscopy of the core CuPc electronic states employing the X-ray excitation (XPS - X-ray Photoelectron Spectroscopy) has been extensively used to characterize stoichiometric CuPc [4-6], only recently a consensus has been reached regarding the correct interpretation of the chemical shifts observed in the C1s composite XPS signal [5,6]. The presence of organic residues may complicate identification of the C1s components. This report presents a systematic XPS study of a large number of commercial CuPc materials and, in particular, the way of resolving the C1s signal into its constituents with the help of corresponding IR absorption data.

EXPERIMENTAL DETAILS

A number of commercial CuPc materials (20 commercial samples) were analyzed. They were obtained from different vendors in form of nanocrystalline powders. All CuPc samples consisted of β-CuPc nanocrystallites having the shape of elongated prisms with dimensions between 50 nm and 200 nm, as confirmed by the electron microscopy. The results were compared with the corresponding data obtained for a selected thin film obtained by thermal evaporation using CuPc purified by a triple-sublimation. Clean, highly conductive Si wafers were used as substrate on which CuPc samples were deposited. XPS specimens were prepared by depositing droplets of aqueous dispersion of a pigment and then allowing the water to dry out leaving behind a solid film residue. Alternatively, the dry powder was pressed against sticky, conductive tape mounted on the Si substrate until it formed a thick, continuous layer of powder. Identical XPS spectra were obtained for specimens of the same pigment prepared by these two methods.

XPS measurements were conducted using a commercial XPS spectrometer equipped with monochromated Al Kα X-ray source (1486.6 eV) and the energy scale was calibrated using the adventitious C 1s signal (284.8 eV) due to residual hydrocarbons. The positions of individual XPS peaks, their widths and amplitudes were determined by means of a non-linear least-square fitting the experimental data with a mixed Lorentzian-Gaussian function (15:85 ratio) after a Shirley-type background was subtracted. The quality of the fitting process was tested by simultaneously conducting the Pearson's χ^2 test; fitting was repeated until satisfactorily low value (< 1) of the χ^2 was achieved. IR absorption was measured using a commercial research-grade FTIR spectrometer. The IR spectra were obtained using pressed tablets containing CuPc pigment and optically neutral KBr providing transmittance between 10% and 90% throughout the entire measured IR spectrum.

DISCUSSION

The C1s signal of a high purity thin-film CuPc consisted of two peaks corresponding to different carbon bonding configuration within the CuPc molecule, namely the carbon within benzene ring (C1) and carbon in pyrrole ring (C2). In addition, these peaks were accompanied by their shake-up satellite due to π - π* transitions (SC1 and SC2). The best fit of binding energies for all analyzed samples yielded the following results: C1 = 284.80 eV, C2 = 286.15 eV, SC1 = 286.61 eV, and SC2 = 288.17 eV. Their respective magnitudes (area under the peak) always satisfied the condition C1 / C2 ratio equal 3 (number of carbon atoms in benzene ring of a CuPc molecule is three times larger than the number of carbon atoms in pyrrole configuration). In addition, the satellite signal was always within 10% to 15% of the corresponding main transition signal, in agreement with the recently calculated transition probabilities [6,7].

In the case of commercial CuPc materials the C1s spectrum could not be fitted with these four peaks. However, detailed analysis showed that satisfactory fit was always obtained when one additional peak, having a binding energy equal 285.68 eV, was added. This additional peak was due to carbon associated with organic impurities present in the commercial CuPc materials (non-CuPc carbon). It is worth noticing that with the introduction of this additional carbon component all previously mentioned conditions imposed on the C1s signal, namely C1/C2 ≈ 3 and SC1/C1 ≈ SC2/C2 ≈ 0.1 to 0.15, were satisfied for all measured CuPc samples.

The origin of the non-CuPc carbon was further elucidated by comparing IR spectra of as-received CuPc materials. Figure 1 shows IR absorption within the spectral range corresponding to C-H vibrations. The observed signal can be divided into two regions: aromatic C-H above 3000 cm^{-1} and aliphatic C-H below 3000 cm^{-1}. The aromatic part remained the same for all samples indicating that it mostly originated from C-H bonds in the peripheral benzene rings of CuPc molecules. Aliphatic C-H signal was due to aliphatic impurities. It varied between the samples indicating different amounts of these impurities present in as-received CuPc materials. Figure 2 demonstrates an excellent agreement between the non-CuPc C1s XPS signal and the aliphatic C-H vibrations. It shows that majority of impurities within the commercial CuPc materials contains large amounts of aliphatic carbon moieties.

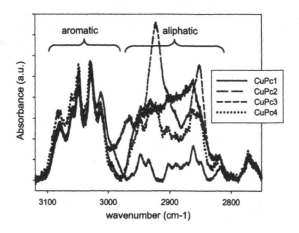

Figure 1. C-H vibration range of the IR absorption spectra for several exemplary commercial CuPc materials. Spectra have been normalized using the pyrrole stretching at 1421 cm^{-1}.

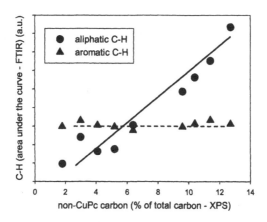

Figure 2. Correlation between the non-CuPc carbon (XPS) and C-C aliphatic / aromatic absorption (FTIR).

CONCLUSIONS

Detailed analysis of the XPS C1s signal provided a very accurate measurement of binding energies of carbon residing in benzene and pyrrole configurations within the CuPc molecule. In addition, binding energies of the respective shake-up states were determined and previously reported relationship between the main and satellite π - π^* transitions was confirmed. The concentration of the carbon residing in pigment additives (non-CuPc carbon) was quantified and the result agreed with the relative concentration of the aliphatic C-H moieties obtained from the FTIR measurement. This work may provide a methodology for routine FTIR evaluation of the organic impurities present in commercial CuPc materials used to fabricate organic electronic devices.

REFERENCES

1. R.H. Friend, R.W. Gymer, A.B. Holmes, J.H. Burroughes, R.N. Marks, C. Tatiani, D.D.C. Bradley, D.A. Santos, J.L. Bredas, M. Logdlund, W.R. Salanek, *Nature* **397**, 121 (1999).
2. Z. Bao, A.J. Lovinger, A. Dodabaladur, *Appl.Phys.Lett.***69**, 3066 (1996).
3. H. Zollinger, *Color Chemistry*, Wiley-VCH & VHCA, Zürich 2003.
4. .T. Schwieger, H. Peisert, M.S. Golden, M. Knupfer, J. Fink, *Phys.Rev.***B66**, 155207 (2002).
5. K. Peisert, M. Knupfer, J. Fink, *Surf.Sci.***515**, 491 (2002).
6. B. Brena, Y. Luo, M. Nyberg, S. Carniato, *Phys.Rev.***B70**, 195214 (2004).
7. B. Brena, S. Carniato, *J.Chem.Phys.***122**, 184316 (2005).

Mater. Res. Soc. Symp. Proc. Vol. 1270 © 2010 Materials Research Society   1270-HH14-40

New Conducting And Semiconducting Polymers For Organic Photovoltaics

Shawn Sapp and Silvia Luebben*
TDA Research, Inc. Wheat Ridge, Colorado 80033

ABSTRACT

In the emerging field of low-cost printed electronics there is a lack of solvent processable conducting and semiconducting materials with highly tuned and known electronic properties. Currently the best performing conductors and semiconductors are not sufficient to produce truly printable, cost competitive organic photovoltaics (OPVs). TDA Research, Inc. (TDA) has been investigating a new class of solvent processable intrinsically conducting polymers for use as charge transport and transparent conducting layers in organic electronic devices. We have also begun the manufacture of electron-deficient semiconducting polymers that may prove to be excellent acceptors in bulk hetero-junction OPVs. This paper presents a summary of the materials characterization conducted on TDA's new electronic materials and how these may address several of the pressing issues preventing the realization of low-cost, printed solar cells and flexible electronics devices.

INTRODUCTION

ICPs are polymers with extended pi conjugation along the molecular backbone, and their conductivity can be changed by several orders of magnitude from a semiconducting state to a metallic state by doping. P-doping is achieved by partial oxidation of the polymer by a chemical oxidant or an electrochemical method, and causes depopulation of the bonding pi orbital (HOMO) with the formation of "holes" [1].

Despite the promises of ICPs since their discovery in the 70s, relatively few commercial products have succeeded, primarily because of their limited performance and inherent insolubility, which makes processing difficult. PEDOT is one of the most commonly used ICPs because of its good electrical conductivity, environmental stability in the doped (conducting) form, and reasonable optical transparency when used as a thin film [2]. A common way to apply a PEDOT coating is to use a water dispersion consisting of a blend of PEDOT and the polyanion poly(styrene sulfonate) or PEDOT-PSS. Doped PEDOT-PSS blends are manufactured by H.C. Starck and are marketed under the trade name of Clevios® P. Several grades of Clevios® P are available from Sigma-Aldrich (655201, 483095, 560596). A low-conductivity grade has been successfully employed as the hole injection layer in organic LEDs (OLEDs) and OPVs and the high conductivity grades are being evaluated as transparent conductors with work functions of ca. 5.1 eV [3,4].

DISCUSSION

Despite the success of the PEDOT-PSS blends, it has been shown that the presence of the strongly acidic and hygroscopic PSS can sometimes degrade device lifetime and performance [5-7]. With recent advances in flexible, printed electronic devices, there is increasing interest in optically transparent conducting polymer materials that can be processed from non-hygroscopic solvents and that will wet hydrophobic plastic substrates. TDA Research, Inc. (TDA) developed and manufactures solvent-dispersible forms of PEDOT. Selected grades of these materials are available through Sigma-Aldrich under the trademark Aedotron[TM] materials. Our approach is to synthesize block copolymers of doped PEDOT and a flexible, soluble polymer such as poly(ethylene glycol) (PEG) [8]. We have developed a number of block copolymer geometries (Fig. 1) and found that through careful control of block composition, molecular weight, block ratio, and dopant type, we can vary the bulk conductivity of the copolymers from 10^{-4} S/cm to 60 S/cm. Fig. 1 shows the chemical structure of both a multi-block and tri-block PEDOT-PEG copolymer. We have developed methods to purify and process our copolymers to form stable colloidal dispersions in organic solvents. These dispersions are neither acidic nor corrosive, and can be used to spin cast or otherwise apply non-hygroscopic thin films of the copolymers on a variety of inorganic and organic substrates. These colloidal dispersions are stabilized by the highly solvated PEG chains which sterically limit the aggregation of the PEDOT blocks. Our copolymers easily disperse in polar aprotic solvents; we have selected propylene carbonate for applications that require a high boiling solvent, and nitromethane for applications that require a volatile solvent. Other solvents are being explored, especially for the low conductivity materials.

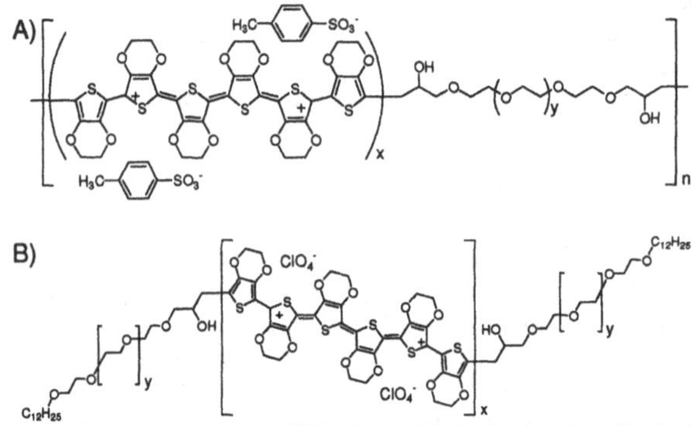

Fig. 1. Chemical structure of TDA's multi-block (a) and tri-block (b) PEDOT-PEG block copolymers.

Since the colloidal stabilization mechanism in our products is independent from the polymer doping, the dopant can be controllably varied to tune the bulk conductivity and the work function of our copolymers. Typically, *para*-toluenesulphonate (PTS) doped copolymers (Aedotron[TM] P polymers) have a lower conductivity, making them useful for antistatic dissipation applications and as an electrode interface layer in OLEDs. Aedotron[TM] P polymers typically have larger particle sizes in suspension and are somewhat amenable to being dispersed in less polar solvents. Perchlorate-doped copolymers (Aedotron[TM] C polymers) typically have a higher conductivity with thin films that are more transparent. With the improved tri-block copolymer (Aedotron[TM] C-3 polymer, Fig. 2B) we can spin cast thin films with 1000 Ohms/square sheet resistance at 80% transmittance (400-800 nm average) with good wetting properties on polycarbonate and other plastic films. Fig. 2 shows the UV-vis spectra for 1, 2, and 3-layer films spun on glass at 1000 RPM, and each data trace is labeled with the measured sheet resistance for that film. These properties meet requirements for a transparent conductor that can be used in touch sensitive displays and electroluminescent lamps and displays. The tri-block copolymer has smaller particle size (290 nm) in suspension than our multi-block copolymers, and form thin films with lower surface roughness (<10 nm), as determined by contact-mode Atomic Force Microscopy.

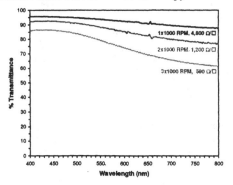

Fig. 2. UV-visible spectra of TDA's new, high-conductivity, tri-block copolymer spin cast at 1000 RPM; 1, 2, & 3-layer films are shown and labeled with the corresponding sheet resistance.

Relative electronic band energies of materials in different layers are important to consider in designing multilayer devices. The work functions of our multi-block copolymers were measured using x-ray photoelectron spectroscopy and were found to be lower than the work function of PEDOT-PSS blends (~4.2 eV for Aedotron[TM] P polymers and 4.3 eV for Aedotron[TM] C polymers) [9]. This lower work function must be taken into account when fabricating thin film electronic devices in which the alignment or overlap of electronic bands is crucial. Table 1 summarizes the properties of the commercial grades of our Aedotron™ copolymers that are commercially available.

Table 1. Comparative Table of TDA's Conducting Polymer Products.

	Aedotron™ C-NM	Aedotron™ C-PC	Aedotron™ P-NM	Aedotron™ C3-NM
	Perchlorate-doped multiblock copolymer in nitromethane	Perchlorate-doped multiblock copolymer in propylene carbonate	PTS-doped multiblock copolymer in nitromethane	Perchlorate-doped triblock copolymer in nitromethane
Bulk conductivity (S/cm)	0.1-2	0.1-2	10^{-3}-10^{-4}	10-60
Sheet resistance, Spin Cast Thin Films* (Ohms/square)	10^{4}-10^{5}			600-3000
Average transmittance** (%T)	70-85%			70%-85%
Particle size in suspension (nm)	600-1000			200-600
RMS roughness spin cast thin films (nm)	40			10
Work Function (eV)	4.33		4.19	
Main Characteristics	General purpose, moderate conductivity dispersion in volatile solvent	General purpose, moderate conductivity dispersion in low-volatility solvent	Hole Injection Layers and low conductivity applications	High transparency and high conductivity dispersion in volatile solvent

* Typically 1-3 layers spun at 1000 RPM or higher
** %T averaged from 400-800 nm, background to Corning Glass

New Polymeric Acceptors

The unifying basic requirement of most thin-film, organic electronic devices like OLEDs and OPVs is that they contain at least two semiconducting materials with offsets in their molecular orbital (HOMO-LUMO) energetic levels. In the organic semiconductor world, one can create such an energy offset by forming an interface between a more electron-rich (donor) semiconductor and an electron-poor (acceptor) material. It is at this interface where charge separation or recombination typically occurs. Moreover, the extent of the offset and the proper alignment of the HOMO-LUMO bands of the donor and acceptor semiconductors are critical to the efficient operation of the device. It is therefore important to have a wide variety of donor and acceptor materials to choose from. There are a number of available classes of relatively electron-rich, semiconducting donor molecules and polymers. In contrast, there are few electron-poor, semiconducting acceptor molecules, like metalloporphyrins and methanofullerenes. Even rarer are the semiconducting, pi conjugated acceptor polymers like cyano-derivatives of poly(p-phenylenevinylenes).

Our group has been working to develop and produce new semiconducting, pi conjugated, acceptor polymers and oligomers. Our approach is quite similar to what has been done for many years to produce electron-rich, donor conducting polymers; we introduce a heteroatom to the pi conjugated backbone that can alter the electron density of the overall polymer. The heteroatom that we add is boron, whose vacant p orbitals are conjugated to the pi electronic system of unsaturated repeat units of the polymer. Because of the absence of electrons in the boron p orbitals, the overall pi electronic system of the polymer becomes inherently electron-deficient and, therefore, the polymer has acceptor-like electronic properties.

Several different synthetic methods are available to prepare air stable pi conjugated organoboron polymers [10-15]. Our group has prepared a number of both new and previously reported pi conjugated organoboron polymers and oligomers. Fig. 3 shows representative chemical structures of the polymers we have been studying. Over the past few years we have refined their synthesis and purification, characterized their properties as organic semiconductors, and evaluated their performance in thin film devices [16]. TDA just begun manufacturing selected pi conjugated organoboron polymers and oligomers under the trade name of Boramer[TM] materials, which are available at Sigma-Aldrich. Additional pi conjugated organoboron structures are currently under development and investigation with funding from NASA.

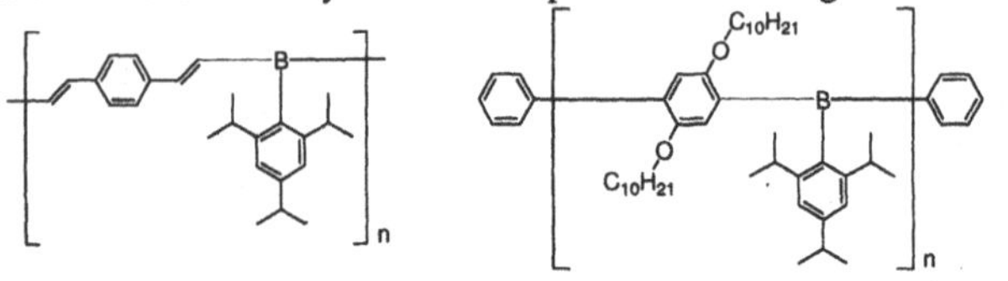

Fig. 3. Chemical structure of TDA's boron-containing, n-type polymers.

We have found that careful purification of the polymers is critical to preserve the solubility of these materials. Chloroform and chlorobenzene are preferred solvents for most of these polymers. All the prepared organoboron polymers are colored and the majority are strongly photoluminescent in the blue to green region of the visible spectrum (see Fig. 4). Air-stability has not been fully assessed yet, but preliminary evidence indicates that it varies with the polymer structure. Boramer[TM] T01 polymer is more sensitive to air than Boramer[TM] TC03 polymer. Since we have yet to quantify their air-stability, we prefer to handle both materials under an inert atmosphere.

Fig. 4. Chloroform solutions of TDA's boron-containing, n-type polymers under ambient (a) and ultraviolet (b) lighting.

The electronic band structure of selected Boramer™ materials was characterized via ultraviolet photoelectron spectroscopy (UPS) at Colorado State University (Fort Collins, CO). UPS gives a direct measure of the electron energies of the HOMO level to Fermi level gap at the low binding energy end of the spectrum. UPS results conclusively prove that our polymers are in fact acceptor semiconductors and that their valence band (VB) resides at a similar energy to the VB of other common acceptor organic semiconductors including methanofullerenes (PCBM) and cyano-PPV. The polymer bandgap was estimated from UV-Vis spectra and was found to be in the range of 2.6-2.9 eV. Fig. 5 shows the HOMO-LUMO levels of two of our boron-containing, pi conjugated polymers (orange) along with familiar donor (blue) and other acceptor organic materials (green). The energy level data clearly indicate their acceptor-like character, with Boramer™ T01 having the lowest lying work function and HOMO-LUMO levels.

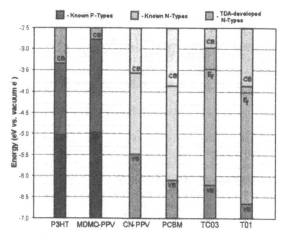

Fig. 5. Energy level diagram of HOMO-LUMO levels for known donor and acceptor organic semiconducting materials, including TDA's new acceptor materials.

During our work on these new acceptor materials we collaborated with the National Renewable Energy Laboratory (NREL, Golden, CO) to evaluate the properties of our polymers for use in OPV prototypes. NREL carried out photoluminescence quenching (PLQ) experiments and built OPV devices with two of the materials supplied by TDA. Results from PLQ indicated that our polymers efficiently quench the excited state of a typical donor semiconductor (MDMO-PPV) with efficiencies up to 83%. This indicates that, in fact, efficient electron transfer occurs from this donor semiconductor to our material. One of the prepared polymers was also used as the electron-transmitting and light–emitting layer for the fabrication of an OLED prototype. Bright green light emission was observed (similar in color to the solid-state luminescence of our polymer) at a turn-on voltage of ca. 6 Volts.

ACKNOWLEDGEMENTS

Many of our coworkers here at TDA have contributed to these efforts so we would like to acknowledge: Emily Chang, Raechelle D'Sa, Cory Kruetzer, and Carolina Wilson for their hard work and diligence. We would like to thank our collaborators at the Univ. of Missouri-K.C., CSU, and NREL; all have been instrumental in characterizing and demonstrating proof-of-concept for our new materials. This work was carried out in part with funding from the NSF (contracts DMI-0319320, OII-0539625, DMI-0110105), NASA contract NNX08CB47P, and OSD (contract N00164-06-C-6042).

REFERENCES

[1] MacDiarmid, A.G. *Angew. Chem. Int. Ed.*, **40**, 2581-2590, 2001.

[2] Groenendaal, L; Dhaen, J.; Manca, J.; Van Luppen, J.; Verdonck, E.; Louwet, F.; Leenders, L. *Synth. Met.*, **135-136**, 115-117, 2003.

[3] Granström, M.; Petritsch, K.; Arias, A. C.; Lux, A.; Andersson, M. R.; Friend, R. H. *Nature*, **395**, 257-260, **1998**.

[4] Cao, Y.; Yu, G.; Zhang, C.; Menon, R.; Heeger, A.J. *Synth. Met.*, **87**, 171-174, 2003.

[5] Kawano, K.; Pacios, R.; Poplavskyy, D.; Nelson, J.; Bradley, D.; Durrant, J. R. *Solar Energy Materials & Solar Cells*, **90**, 3520-3530, 2006.

[6] Danier Van Der Gon, A.W.; Birgerson, J.; Fahlman, M.; Salaneck, W.R. *Org. Electr.*, **3**, 111-118, 2002.

[7] Greczynski, G.; Kugler, T.; Keil, M.; Osikowicz, W.; Fahlman, M.; Salaneck, W. R. *J. Elec. Spectr. & Rel. Phenom.*, **121**, 1-17, 2001.

[8] Luebben, S.; Elliott, B.; Wilson, C. "Poly(heteroaromatic) Block Copolymers with Electrical Conductivity," U.S. Patent 7,361,728 B1 Issued on April 22, 2008.

[9] Sapp, S.; Luebben, S.;Jeppson, P.; Shulz, D.L.; Caruso, A.N. *Appl. Phys. Lett.*, **88**, 152107(1-3), 2006.

[10] Chujo, Y.; Miyata, M.; Matsumi, N. *Polymer Bulletin*, **42**, 505-510, 1999.

[11] Chujo, Y.; Miyata, M.; Matsumi, N. *Macromolecules*, **32**, 4467-4469, 1999.

[12] Chujo, Y.; Naka, K.; Matsumi, N. *J. Am. Chem. Soc.*, **120**, 10776-10777, 1998.

[13] Chujo, Y.; Umeyama, T.; Matsumi, N. *Polymer Bulletin*, **44**, 431-436, 2002.

[14] Jäkle, F.; Sundararaman, A.; Victor, M.; Varughese, R. *J. Am. Chem. Soc.*, **127**, 13748-13749, 2005.

[15] Parab, K.; Venkatasubbaiah, K.; Jäkle, F. *J. Am. Chem. Soc.*, **128**, 12879-12885, 2006.

[16] Luebben, S.; Sapp, S.A. "Use of Pi-conjugated Organoboron Polymers in Thin-film Organic-Polymer Electronic Devices" PCT Patent Application PCT/US07/64328.

Mater. Res. Soc. Symp. Proc. Vol. 1270 © 2010 Materials Research Society   1270-HH14-54

Tanninsulfonic Acid Doped Polyaniline as a Novel Heterojunction Material in Hybrid Organic/Inorganic Solar Cells

Shawn E. Bourdo[1,2], Viney Saini[1], Brock A. Warford[2], Florent Prou[1,3], Venugopal Bairi[2], Zhongrui Li[1], A. S. Biris[1], Tito Viswanathan[2]

[1]Nanotechnology Center, University of Arkansas at Little Rock, Little Rock, AR 72204, U.S.A.
[2]Department of Chemistry, University of Arkansas at Little Rock, Little Rock, AR 72204, U.S.A.
[3]Ecole d'Ingenieurs du CESI-EIA, La Couronne, France

ABSTRACT

Polyaniline (PANI) is a widely researched inherently conducting polymer (ICP) with applications in organic photovoltaic technology such as hole-transport layers. Less work has been reported on its use as an active layer in heterojunction solar cells. We report here that tanninsulfonic acid, a renewable resource, is a photoactive macromolecule that has been incorporated into the polyaniline chain. The presence of tanninsulfonic acid has been confirmed by infrared spectroscopy. The product has been processed into solutions using camphorsulfonic acid and m-cresol from which films are cast.

Tanninsulfonic acid-doped polyaniline, TANIPANI, is an attractive ICP for use in low-cost PV devices due to a facile synthesis and its stability in air. Similar to other ICPs, this material can be processed by traditional techniques such as spin-coating, air-brushing, ink-jet printing, etc. that allow it to be easily scalable. The structural characteristics of TANIPANI are analyzed by UV-Vis and infrared spectroscopy.

TANIPANI has been incorporated into a hybrid organic/inorganic photovoltaic device for the first time. Devices were fabricated by spin-coating solutions of TANIPANI onto n-type silicon (n-Si) substrates. The illuminated and dark current-voltage characteristics of devices are reported. For the devices produced, the maximum short circuit current, open circuit potential, fill factor and power conversion efficiency were 11 mA/cm^2, 0.430 V, 0.24, and 1.11%, respectively.

INTRODUCTION

Photovoltaic research has become increasingly popular across many materials disciplines. The necessity for cheaper routes to renewable energy devices and increase in device efficiencies are two driving forces for the interest. Organic materials have been looked upon as a potential to drive down the production costs due to ease of processing by techniques such as ink-jet printing and spin-coating[1]. Much of the focus of materials for organic photovoltaics (OPV) has been applied to either bulk heterojunction devices or a dye-sensitized solar cells (DSSC)[2]. While hybrid organic/inorganic devices may not be as attractive as the aforementioned solar cells due to the cost of the inorganic component, some fundamental questions can be answered when exploring new organic materials for PV technologies. While polyaniline has been touted for use as a hole transport layer in heterojunction OPVs and DSSCs[3-5], there have also been reports on PANI/n-Si p-n heterojunction solar cells with promising results[6, 7]. It is our hope that exploration of some novel polyaniline-based materials will lead to exciting results for many types of OPVs being explored in our labs.

A variety of metal and metal-oxide nanoparticles can be incorporated into our polymer network by using tanninsulfonates. Tannins are naturally occurring polyphenols that are found in the vascular tissue of plants such as the leaves, bark, grasses, and flowers. Tannins are classified into two groups [8]: condensed tannins or proanthocyanidins and hydrolysable tannins from the polyesters of gallic acids.

Figure 1 illustrates the reaction scheme for the sulfonation of a monomeric unit of condensed tannin that we have utilized in this research. The structure consists of three rings: two benzene rings on either side of an oxygen-containing heterocyclic ring. The A-ring to the left of the cyclic ether ring consists of one or two hydroxyl groups. The B-ring present on the right of the cyclic ether ring also consists of two or three hydroxyl groups.

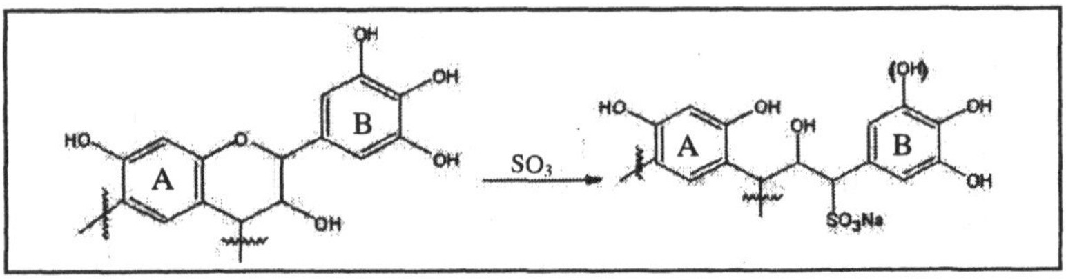

Figure 1. Reaction scheme for the sulfonation of a monomeric unit of a condensed tannin.

The sulfonation of tannin allows for this macromolecule to be incorporated into polyaniline as a dopant (sulfonate) as well as being covalently grafted to the polymer chain via the aromatic substitution. The o-dihydroxy phenyl (catechol) groups in the tannin molecule are responsible for the chelation properties of the tannins. So tannin in this preparation can also be used for binding the metal oxides of which we are interested for potential photovoltaic properties. We have tried to incorporate titanium (IV) oxide, aluminum oxide, and zinc oxide into our products and have evaluated them in hybrid organic/inorganic solar cells.

EXPERIMENTAL

Synthetic methods

Synthesis of tanninsulfonic acid-doped polyaniline (TANIPANI) was accomplished using a biphasic system made up of aqueous and organic solvents. An aliquot (120mL) of 1M methane sulfonic acid (HMSA) and was added to a jacketed beaker followed by a 0.1g sample of tanninsulfonate. In some cases a 0.05g sample of metal oxide nanopowders [TiO_2 (~21nm particle size), Al_2O_3 (average 13 nm particle size), ZnO-6% Al doped (<50nm) all obtained from Aldrich] was added to the solution at this time. Two milliliters of aniline was then added to the above solution followed by addition of 80ml of isopropyl alcohol. The reaction solution was cooled between -10°C and -15°C. A sample of sodium persulfate (1.12g) in 40 ml of 1M HMSA was chilled and added to the reaction mixture dropwise. The reaction was allowed to continue for one day after the reaction turned green in color. The resulting product was filtered, washed with deionized water until the filtrate was clear, and then a final wash with 1M NH_4OH.

The product was fully dedoped by stirring in 200ml of the 0.1M ammonium hydroxide solution overnight. The product was filtered and washed with deionized water until the filtrate was at neutral pH. The obtained TANIPANI base was vaccum dried and stored for further use. The products were characterized by FTIR & UV/Visible spectroscopy and X-ray diffraction.

Solar cell device fabrication

A solution of doped TANIPANI (or TANIPANI-metal oxide composite) solution was prepared by mixing the undoped TANIPANI (or TANIPANI-metal oxide composite) with camphorsulfonic acid (HCSA), adding m-cresol, and then homogenizing for 30 minutes; the solution was allowed to stir overnight in a closed vessel. The solution was spun onto RCA cleaned n-Si wafers (1.5 cm x 1.5 cm) at an initial rate of 250 rpm for 30 seconds and then at 800rpm for 5 minutes; films were also spun onto quartz slides for spectral characterization. All films were cured in an oven at ~70°C for at least one hour and then pressure was reduced in order to remove any residual solvent. Silver was deposited using a Gatan Ion-beam coater (conditions: 5keV and 150μA for 5 minutes). Electrical contacts were made directly to the top (TANIPANI) and bottom of the device (n-Si) using bulldog clips. A mask was placed over the film so that an area of the device would be exposed that had a uniform TANIPANI film.

Characterization methods

A Nicolet MAGNA-IR 550 Series 2 Spectrometer was used to analyze the FTIR spectra of the powder samples as pressed pellets in KBr. A Perkin-Elmer UV/Vis/NIR Lambda 19 Spectrometer was used to analyze the ultraviolet and visible spectrum of undoped samples in N-methylpyrrolidinone and HCSA-doped samples as thin films. X-ray diffraction patterns of undoped TANIPANI and TANIPANI-metal oxide composite powder samples were measured on a Bruker AXS D8 Discover with a GADDS 2D counter. Copper K-alpha line was used as excitation source; X-ray tube runs at 40 kV and 35 mA. The current-voltage (IV) characteristics of solar cell devices were tested under dark and illuminated conditions (1.5 AM, 100mW/cm^2) using a Keithley 2400 Sourcemeter.

RESULTS AND DISCUSSION

Characterization of TANIPANI

After the synthesis of TANIPANI and metal oxide composites, the products were characterized to determine the presence of tanninsulfonate and metal oxide prior to solution processing. Infrared spectroscopy was used to determine if a successful polymerization had occurred. Several characteristic peaks exhibited by polyaniline are present in our TANIPANI samples. The vibrational mode around 3400 cm^{-1} results from the N-H bond, while peaks at 1600 cm^{-1} and 1500 cm^{-1} result from the quinoid and benzenoid vibrations, respectively[9]. Previous work has been performed by a colleague identifying the peak at 1030 cm^{-1} as a way of proving incorporation of tanninsulfonate into the final product[10]. The infrared spectra of the undoped samples are presented in Figure 2 (left) along a dashed line placed at 1030 cm^{-1} for emphasis of the sulfonate stretch.

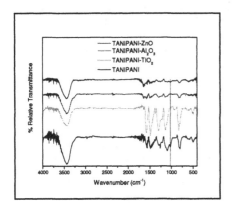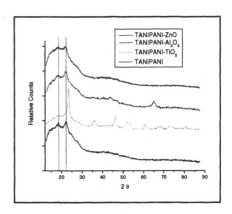

Figure 2. Infrared spectra (left) and X-ray diffraction patterns (right) of TANIPANI, TANIPANI-TiO₂, TANIPANI-Al₂O₃, and TANIPANI-ZnO [from bottom to top].

Since the TANIPANI is in an undoped form after the synthesis, by treating with base, the observation of the sulfonate stretch is a result of the tannin being grafted onto the polyaniline backbone. In this manner tanninsulfonate functions not only as a dopant as the name tanninsulfonic acid-doped polyaniline implies, but is covalently bound to the polymer and cannot be simply washed away as in normal protonation-deprotonation of dopants. The infrared spectra did not indicate any changes from the incorporation of the metal oxide into the synthesis and therefore further characterization was performed.

X-ray diffraction was performed on the powder samples to detect the presence of metal oxides remaining in our samples prior to solution processing. The main features of the TANIPANI pattern is a broad diffraction from about 12° – 30° 2θ as seen above in Figure 2 (right) along with two diffractions at 18.5°and 22.0° 2θ that result from stacking of the polymer chains[11, 12]. These two diffractions are dashed for comparison among the 4 patterns. Among the other diffraction patterns TANIPANI-TiO₂ and TANIPANI-Al₂O₃ samples have several features in their diffraction pattern that have been attributed to the tannin complexed metal oxides from the respective synthesis.

Solar cell characterization

The undoped TANIPANI and composites were then processed into conductive solutions to be used as layers in heterojunction devices. It must be determined that the form of TANIPANI that we are using for our devices is the doped form giving rise to polaronic/bipolaronic transitions. When polyaniline becomes doped, optical absorptions occur at around 350 nm, 440 nm, and > 600 nm that correspond to $\pi-\pi^*$, polaron-π^*, and π-conduction band electronic transitions, respectively [13]. These classical transitions for a sample of doped polyaniline are observed for all samples; therefore we can be confident that each film is in the conductive form of polyaniline.

Several solar cell devices were fabricated using a spin-coating technique. As our fabrication process has progressed we have started to use masks to expose specific areas of the devices to illumination. The size of masks was varied to gain a better understanding of the relationship between device area and characteristics throughout these studies. The masks were approximately 5mm x 5mm and 3mm x 3mm. In the IV curves displayed in Figure 3, the masks are denoted by sizes (i.e. 5x5, 3x3, and 1x1) and conditions (i.e. light or dark).

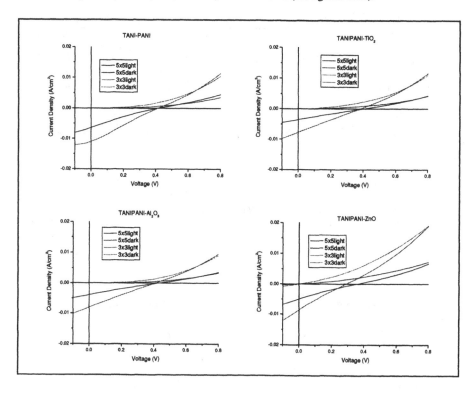

Figure 3. Current voltage characteristics of devices fabricated from TANIPANI (top left), TANIPANI-TiO$_2$ (top right), TANIPANI-Al$_2$O$_3$ (bottom left), and TANIPANI-ZnO (bottom right)

The overall results are promising since these results are the first such use of TANIPANI in an energy related device. The use of TANIPANI alone was a success, and when some metal oxide nanopowders were integrated in the polymer samples, devices continued to show good response. When the solar cells are illuminated, the photoexcited charge carriers at the n-Si/PANI heterojunction dissociate into free electrons and holes. These free electrons are transported through n-Si to the cathode and holes are swept away to anode through the PANI layer. For all the devices, a smaller area exposed to light yielded much better efficiency, fill factor (FF), and short-circuit current density (J$_{SC}$) than when a larger area exposed. The open circuit voltage

(V_{OC}) was the only parameter that decreased as a result of the smaller mask size. Table 1 summarizes characteristic device parameters obtained from the IV curves.

Table 1. Overall characteristics for solar cell devices.

Sample (mask)	Mask area (cm²)	efficiency (%)	Fill Factor	V_{OC} (V)	J_{SC} (mA/cm²)
TANIPANI (5x5)	0.27	0.56	0.21	0.418	6.39
(3x3)	0.10	1.11	0.24	0.418	11
TANIPANI-TiO₂ (5x5)	0.27	0.33	0.24	0.402	3.51
(3x3)	0.10	0.71	0.24	0.381	7.69
TANIPANI-Al₂O₃ (5x5)	0.27	0.38	0.23	0.430	3.88
(3x3)	0.10	0.79	0.24	0.410	8.07
TANIPANI-ZnO (5x5)	0.27	0.38	0.22	0.358	4.97
(3x3)	0.10	0.57	0.23	0.290	8.59

CONCLUSIONS

We have shown here for the first time the use of TANIPANI as a potential heterojunction material in hybrid solar cells with device efficiencies up to 1.11%. This represents a crucial step towards the utilization of this conducting polymer in other PV devices also. While we have shown here that it is possible to incorporate these nanoparticles into TANIPANI and fabricate a PV device from the product, there is still much work to be done with respect to the interaction of the nanoparticles in the overall device.

ACKNOWLEDGEMENTS

We gratefully acknowledge support from the U.S. Department of Energy (Grant No. DE-FG 36-06 GO 86072).

REFERENCES

1.   C.J. Brabec and J.R. Durrant, MRS Bull. **33,** 670-675 (2008).
2.   H. Hoppe and N.S. Sariciftci, J. Mat. Res. **19**(7), 1924-1945 (2004).
3.   H. Bejbouji, et al., Sol. Energy Mater. Sol Cells, **94**(2), 176-181 (2009).
4.   S. Tan, et al., Langmuir, **20,** 2934-2937 (2004).
5.   R. Valaski, et al., J. of Solid State Electrochem., **10,** 24-27 (2006).
6.   W. Wang and E.A. Schiff, App. Phys. Lett., **91,** 133504 (2007).
7.   W. Wang, E. Schiff, and Q. Wang, J. of Non-Cryst. Solids, **354,** 2862-2865 (2008).
8.   N. Vivas, et al., J. Sci. Food Agric., **72,** 309-317 (1996).
9.   Quillard, S., et al., Phys. Rev. B, **50**(17), 12496-12508 (1996).
10.   K. Taylor, PhD Dissertation, University of Arkansas at Little Rock, 2006.
11.   M.E. Jozefowicz, et al., Phys. Rev. B, **39**(17), 12958-61 (1989).
12.   J.P. Pouget, et al., Macromolecules **24,** 779-789 (1991).
13.   D.M. Tigelaar, et al., Chem. of Mat., **14,** 1430-1438 (2002).

Mater. Res. Soc. Symp. Proc. Vol. 1270 © 2010 Materials Research Society 1270-HH14-56

Modeling the Effect of Annealing and Regioregularity on Electron and Hole Transport Characteristics of Bulk Heterojunction Organic Photovoltaic Devices

Shabnam Shambayati[1], Bobak Gholamkhass[2], Soheil Ebadian[1], Steven Holdcroft[2], and Peyman Servati[1]

[1]Electrical and Computer Engineering Department, University of British Columbia, Vancouver, BC V6T 1Z4, Canada

[2]Department of Chemistry, Simon Fraser University, Burnaby, BC V5A 1S6, Canada

ABSTRACT

In this study, the dark current-voltage characteristics of electron-only and hole-only poly(3-hexyl thiophene) (P3HT):[6,6]-phenyl C_{61}-butyric acid methyl ester (PCBM) as a function of regioregularity (RR) and annealing time is investigated using the mobility edge (ME) model. This model is used to analyze the degradation of electron and hole mobilities as a function of annealing time for 93%-RR and 98%-RR P3HT:PCBM devices. The hole mobility is almost unchanged by the RR nature of P3HT and thermal annealing. The electron mobility, however, behaves differently after annealing. The electron mobility of 98%-RR devices, which is initially higher than that of the 93%-RR devices, experiences a steep decline with annealing. Based on ME analysis, this is due to an increase in trap states in the exponential tail caused by phase segregation of solid state blends of 98%-RR polymer and PCBM. The electron mobility of 93%-RR devices increases with annealing due to an optimization of nano-phase separated morphology.

INTRODUCTION

In recent years, organic photovoltaic (OPV) devices have shown significant potential for use as low cost, lightweight and flexible solar cells. Many different OPV structures have been investigated; the most promising candidate is bulk heterojunction (BHJ) solar cells, namely, BHJs based on poly(3-hexyl thiophene) (P3HT):[6,6]-phenyl C_{61} butyric acid methyl ester (PCBM) blends [1]. The electron-donating polymer and electron-accepting fullerene constituents form a nanometer-sized binary network, referred to as a "donor-acceptor (D-A) BHJ". Photogenerated excitons are dissociated into free charge carriers at the junction between the P3HT and PCBM interpenetrating networks, and the bi-continuity of the network provides separate pathways for hole and electron transportation to their respective electrodes [2].

The photocurrent of the OPV devices is influenced by several parameters, most significantly light absorption, charge transport, and charge separation efficiency. One of the most important limiting factors for these solar cells is mobility [3]. It has been shown that mobility can be improved by optimizing the morphology of the P3HT:PCBM networks and the molecular packing [4]. One method of achieving an optimum morphology is through increasing the degree of regioregularity (RR); regioregularity is the percentage of monomers in the head-to-tail configuration rather than head-to-head [3]. Photocurrent has been shown to increase through improved crystallinity of the polymer as a result of a higher RR [3].

The initial enhancement in performance with increased RR, however, will not remain stable during long-term operation or after annealing [5]. P3HT and PCBM are not miscible; over time, blends of the two may segregate, forming discontinuous microscopic domains. Crystallization is reported to exclude PCBM from the ordered polymeric domains in bulk heterojunctions [6]. Crystallization-driven phase separation is thought to be the main cause for a reduction in D-A interfacial area [7].

While the degradation of regioregular OPVs has been previously documented [5], these processes have not been quantitatively modeled. Electron and hole mobilities are significantly influenced by the purity and bi-continuity of the network, which changes with aging [8, 9]. To quantify the electronic properties of the OPV device, a suitable model should include the energy distribution of the carriers, electron and hole mobilities, mobility degradation with respect to annealing time, and the effect of disorder-induced localized states on transport properties. In this work, we use a mobility edge (ME) model, which is widely used for inorganic amorphous and polycrystalline material, as well as organic polymers [10, 11]. In the following sections, after introducing the ME theory, the details of PV, electron-only, and hole-only device fabrication are described. The effects of aging on hole and electron mobility is discussed based on ME model analysis.

THEORY

This model can be used to extract mobility values from the J-V characteristics of a device by considering the trap density in the space charge limited (SCL) region. In a diode consisting of injecting electrodes that sandwich the semiconductor material, the SCL current density can be written as [12]:

$$J = q\mu_b N_b \kappa \left(\frac{\varepsilon}{qN_b} \right)^{\alpha} (V_D - V_t)^{\alpha+1}/t^{2\alpha+1}, \tag{1}$$

where q is the elementary charge, μ_b the carrier mobility in the hopping band, N_b the effective state density for the transport band, ε the dielectric constant, V_D the diode voltage, t the BHJ film thickness, V_t the threshold voltage, and $\kappa = \alpha^{\alpha}(2\alpha + 1)^{(\alpha+1)}/(\alpha + 1)^{2\alpha+1}$. The power parameter, $\alpha = T_t/T$, is a measure of trap density; T_t is the characteristic slope of the exponential localized tail states in the density of states vs. energy curve, where a higher T_t indicates a higher density of traps in the band tail. α is extracted from the power-law J-V characteristics of the device, $J = k(V_D - V_t)^{\alpha+1}$. It follows that $J/g = (V-V_t)/(\alpha+1)$ where $g = \partial J/\partial V$. Therefore, to extract α, the slope of the J/g versus V curve can be used [12].

In order to find mobility, we must determine the relationship between the density of carriers excited to the isoelectronic transport band, n_{band}, and the density of trapped carriers, $n_{trapped}$ [12, 13]. Using that relationship, effective mobility can be written as:

$$J = q\mu_{eff} N_0 \kappa \left(\frac{\varepsilon}{qN_0} \right)^{\alpha} (V_D - V_t)^{\alpha+1}/t^{2\alpha+1}, \tag{2}$$

where μ_{eff} is the effective mobility, and N_0 is a reference carrier concentration equal to 2×10^{16} cm^{-3}.

EXPERIMENT

Hole-only, electron-only, and PV devices were made using 93% and 98%-RR P3HT and PCBM blends with a 1:1 blend ratio. The detailed fabrication method has been reported previously [14]. All devices were made from P3HT:PCBM blends in1,2-dichlorobenzene (DCB). The BHJ film thickness was uniformly 170 ± 10nm for all samples. Top electrodes for hole-only devices were fabricated by evaporating palladium on the devices; the overall configuration was ITO/poly(3,4-ethylenedioxythiophene):poly(4-styrenesulfonate) (PEDOT:PSS)/P3HT:PCBM/Pd. Due to the difference between the work function and the lowest unoccupied molecular orbital (LUMO) of PCBM, electron injection from PCBM is suppressed [4]. Similarly, electron-only devices were fabricated by evaporating an 80 nm aluminum layer on glass, spin-coating the blend, and evaporating another 100 nm of aluminum as the top electrode [15]. All devices were made in a nitrogen-filled glove box, with oxygen levels less than 0.6 ppm. Testing was done with the samples under vacuum; none of the electron-only devices came in contact with oxygen. The devices were annealed at 150°C for up to five hours.

RESULTS

Photovoltaic property. Figure 1 shows the J-V characteristics of 93%-RR and 98%-RR PV devices under AM1.5 illumination before and after annealing. The as-cast 98%-RR devices initially exhibit superior performance, but the current density decreases significantly after thermal annealing. The current density of 93%-RR devices, however, does not change significantly after the annealing, though it is initially lower than that of the 98%-RR devices. To explain this in terms of mobility, we explore the dark current characteristics for hole-only and electron-only devices.

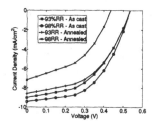

Figure 1: J-V characteristics of devices with different regioregularities, for as cast and annealed conditions.

Hole mobility. Figure 2(a) shows the dark J-V characteristics of the hole-only devices, fabricated using 93% and 98%-RR P3HT:PCBM blends, before and after 2 hours of annealing at 150°C. For the as-cast devices, the 98%-RR device shows slightly larger current than the 93% device. This can be attributed to the higher absorbance in the 98%-RR device due to the superior organization of P3HT domains, enhanced degree of crystallinity, and more efficient electronic conduction. Figure 2(b) shows the J/g vs. V curves for all devices, indicating that conduction is ohmic even before annealing. Even though it is generally expected that annealing increases P3HT hole conduction through improving P3HT crystallinity, the observed improvement is minimal.

This is due to the slower drying time of the solvent (DCB); the solvent-assisted annealing optimizes the crystallinity of the P3HT, even before thermal annealing [16]. Extracted hole mobility is 6.2×10^{-4} cm^{2}V^{-1}s^{-1} and remains unchanged after annealing. We believe that the minor changes in hole mobility cannot explain the main cause of the performance degradation, and therefore, we focus on electron-only devices from this point on.

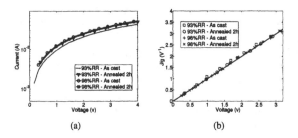

(a)   (b)

Figure 2: (a) J-V characteristics for as-cast and annealed hole-only 93% and 98% RR devices, (b) J/g vs. V, of which the slope m is used for extraction of α, where $m = 1/(1+\alpha)$.

Electron mobility. Figure 3 shows the J/g vs V characteristics for the electron-only (a) 93% and (b) 98% devices. Figure 3(a) demonstrates that α is increasing with annealing time. From comparison of Figure 3(a) and 3(b), it is evident that the increase in α is more pronounced for 98%-RR devices. According to the ME model, the power parameter α is a measure of trap density; the larger the α the higher the density of traps in the band tail. Therefore, a larger increase in α upon thermal annealing indicates a reduction in continuity of the electron carrier phase, i.e., PCBM domains. This can be attributed to the crystallization-driven phase segregation of 98%-RR P3HT, resulting in the crystallization/exclusion of PCBM molecules from ordered domains. Formation of PCBM crystallites generates dead-end; therefore, more traps present in the band tail. The difference in the rate of phase separation process with respect to thermal annealing time and RR of P3HT has previously been reported [17]. In that work it was confirmed that thermal annealing induces the formation of many needle-shape PCBM crystals that are several tens of micrometers in length in the P3HT:PCBM blend films. Specifically, the phase segregation and formation of PCBM crystals are far more extensive in 98%-RR films than in 93%-RR films.

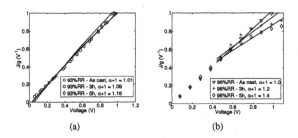

(a)   (b)

Figure 3: J/g vs V characteristics showing α; (a) 93% RR; (b) 98% RR.

Figure 4(a) shows the change in electron mobilities with time as calculated using Equation 2. The electron mobility of the 98%-RR devices, initially larger than the 93%-RR devices, falls rapidly with thermal annealing. This is because annealing causes PCBM to crystallize, producing "clusters" of larger crystals [5]. These clusters reduce phase bi-continuity, causing the electron mobility to decrease, and justifying the increase in α. On the other hand, the electron mobility of 93%-RR devices initially increases with annealing. Thermal annealing improves the development of nano-phase separated domains to an optimum limit. However, further annealing seems to have an insignificant effect on the electron mobility of the 93%-RR devices.

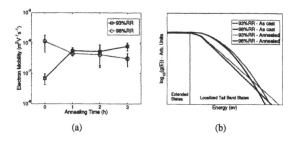

(a) (b)

Figure 4: (a) Extracted mobility values for 93% RR and 98% RR devices over time; (b) Difference between the shape of DOS vs energy graphs for 93%RR and 98% RR.

The observed difference in the voltage dependence of α is used to approximately sketch density of states (DOS) vs. energy, shown in Figure 4(b). Despite the limited accuracy, it can be seen that the band tail in the 98%-RR sample is larger, which indicates the presence of more local traps in that region. Figure 4(b) also shows the effect of annealing on the density of states (DOS): the band tails for both samples are broadened, indicating that more localized traps are present after annealing.

The devices were annealed for only three hours. More experiments are recommended to investigate longer annealing effects on electron mobility.

CONCLUSIONS

We have investigated the effects of annealing and regioregularity on the charge transport characteristics of P3HT:PCBM photovoltaic cells. Initially, the hole mobility of the 98%-RR devices was larger than that of the 93%-RR devices, but after annealing, the current densities became comparable. While a minor enhancement was observed for hole mobilities as a result of annealing, the change in hole mobility was ruled out as a possible cause for performance degradation. The electron mobilities, however, were found to behave differently after annealing. After thermal annealing, mobility decreased in 98-%RR devices, and increased in 93%-RR devices. This is reflected in the power parameter α in the mobility edge model. According to the ME model, the increase in α upon thermal annealing is attributed to an increase in trap states in the exponential tail due to phase segregation in blends of 98%-RR polymer and PCBM. These results show that electron transport plays a critical role in the degradation of effective mobility of P3HT:PCBM solar cells.

ACKNOWLEDGEMENTS

This work was supported by the Natural Sciences and Engineering Research Council of Canada (NSERC) and Canada Foundation for Innovation (CFI). The authors would like to thank Dr. John Madden, Dr. Jaclyn Brusso, and Alexandros Dimopoulos for their guidance during the experimental phase of the project.

REFERENCES

1. A. Baumann, J. Lorrmann, C. Deibel, and V. Dyakonov, *Appl. Phys. Lett.* **93**, 252104 (2008).
2. F. Yang and S.R. Forrest, *ACS Nano* **2**, 1022-1032 (2008).
3. Y. Kim, S. Cook, S.M. Tuladhar, S.A. Choulis, J. Nelson, J.R. Durrant, D.D.C. Bradley, M. Giles, I. McCulloch, C. Ha, and M. Ree, *Nat Mater* **5**, 197-203 (2006).
4. V.D. Mihailetchi, H. Xie, B. deBoer, L. Koster, and P. Blom, *Advanced Functional Materials* **16**, 699-708 (2006).
5. S. Bertho, G. Janssen, T.J. Cleij, B. Conings, W. Moons, A. Gadisa, J. D'Haen, E. Goovaerts, L. Lutsen, J. Manca, and D. Vanderzande, *Solar Energy Materials and Solar Cells* **92**, 753-760 (2008).
6. K. Sivula, Z. Ball, N. Watanabe, and J. Frchet, *Advanced Materials* **18**, 206-210 (2006).
7. K. Sivula, C.K. Luscombe, B.C. Thompson, and J.M.J. Frchet, *J. Am. Chem. Soc* **128**, 13988-13989 (2006).
8. M. Campoy-Quiles, T. Ferenczi, T. Agostinelli, P.G. Etchegoin, Y. Kim, T.D. Anthopoulos, P.N. Stavrinou, D.D.C. Bradley, and J. Nelson, *Nat Mater* **7**, 158-164 (2008).
9. H. Hoppe and N.S. Sariciftci, *J. Mater. Chem.* **16**, 45-61 (2006).
10. A. Salleo, T.W. Chen, A.R. Vlkel, Y. Wu, P. Liu, B.S. Ong, and R.A. Street, *Physical Review* **B70**, 115311 (2004).
11. V. Kumar, S.C. Jain, A.K. Kapoor, J. Poortmans, and R. Mertens, *J. Appl. Phys.* **94**, 1283 (2003).
12. E. Kymakis, P. Servati, P. Tzanetakis, E. Koudoumas, N. Kornilios, I. Rompogiannakis, Y. Franghiadakis, and G.A.J. Amaratunga, *Nanotechnology* **18**, 435702 (2007).
13. P. Servati, A. Nathan, and G. Amaratunga, *Phys. Rev.* **B74**, 245210 (2006).
14. B. Gholamkhass, T.J. Peckham, and S. Holdcroft, *Polym. Chem.*(2010).
15. M. Lenes, M. Morana, C.J. Brabec, and P.W.M. Blom, *Advanced Functional Materials* **19**, 1106-1111 (2009).
16. W. Ma, C. Yang, X. Gong, K. Lee, and A. Heeger, *Advanced Functional Materials* **15**, 1617-1622 (2005).
17. S. Ebadian, B. Gholamkhass, S. Shambayati, S. Holdcroft, and P. Servati, *Solar Energy Materials and Solar Cells* In Press, Corrected Proof (2010).

Mater. Res. Soc. Symp. Proc. Vol. 1270 © 2010 Materials Research Society　　　　　1270-HH14-58

Optical and Electrical Modeling of Polymer Thin-Film Photovoltaics

Wenjun Jiang[1] and Scott T. Dunham[2,1]
[1]Department of Physics, University of Washington, Seattle, WA 98195, U.S.A.
[2]Department of Electrical Engineering, University of Washington, Seattle, WA 98195, U.S.A.

ABSTRACT

　　By coupling optical and electrical device simulation, we explore the optimization of organic photovoltaic devices. We use optical simulation via FDTD and transmission matrix solution of Maxwell's equations in order to calculate distribution of light intensity, including the reflectivity and transmissivity of transparent anode layers. We derive simple formulas for the conductivity requirements as function of pattern structure of potential replacements for ITO as transparent anode material, and explore the optimization of thickness of active layer and its relationship with the peaks of optical generation inside the active layer.

INTRODUCTION

　　Organic semiconductors have attracted a lot of attention due to the flexibility and low weight, as well as ultra-low cost production compared to silicon-based solar cells [1, 2]. Many efforts have been made in order to improve the efficiency of organic solar cells by developing new materials with smaller bandgap, modifying the thickness of active layer, as well as using charge-collecting materials with higher conductivity [3-6]. In this work, we describe efforts at coupled optical and electrical modeling with the aim of optimizing material choices and device structures for maximum performance. The optical processes inside organic solar cells include electromagnetic transmission, reflection, and absorption in multilayer structures. We explored the optimized active layer thickness and choices of using ultra-thin metals to replace ITO as transparent anode layer, with the aim of coupling light strongly into P3HT/PCBM layer while simultaneously reducing cost. From the electrical device point of view, we have derived a simple formula for dependence of required contact sheet resistance on the width and spacing of interdigitated anode contact metal. We use Sentaurus Device simulator (Synopsys), adapted with material-specific models for electronic polymers, to explore impact of illumination and material properties on device behavior. Finally, the coupling of optical and electrical simulation helps us to find the trade-off solutions to device geometry. These properties are then used as the basis of computational exploration of device structure and materials to optimize solar cell performance.

MODELS

Finite-difference time-domain (FDTD) and Transfer matrix method (TMM)

　　Transfer matrix method [7] is used to calculate the propagation of plane waves through layered media. According to Maxwell's equations, the electric field $E(z) = E_r e^{ikz} + E_l e^{-ikz}$ and its first order derivative $F(z) = ikE_r e^{ikz} - ikE_l e^{-ikz}$ across boundaries from one medium to the next can be expressed by simple continuity conditions:

$\begin{pmatrix} E(z+L) \\ F(z+L) \end{pmatrix} = \vec{M} \cdot \begin{pmatrix} E(z) \\ F(z) \end{pmatrix}$, where $\vec{M}$ is the transfer matrix. So the transfer matrix for a system with N layers can be expressed as: $\vec{M_s} = \vec{M_N} \cdots \vec{M_2} \cdot \vec{M_1}$.

　　Finite-difference time-domain method [13-15] discretizes the time-dependent Maxwell's equations using central-difference approximations to the space and time partial derivatives. At any point in space at any time, the updated value of E-field can be expressed as a function of E-field and curl of magnetic field: $\vec{E}(z, t + \Delta t) = F(\vec{E}(z, t), \nabla \times \vec{H}(z, t), t, \Delta t)$. It can deal with electromagnetic wave interactions with any kind of materials and structures as long as materials' permeability, permittivity and conductivity are specified.

Parameter modification for ultra-thin metal layer

The reflectivity and transmissivity equations for a plane parallel, absorbing film situated between two dielectric media can be expressed as [7]:

$$R = \frac{\rho_{12}^2 \exp(2k_2\eta) + \rho_{23}^2 \exp(-2k_2\eta) + 2\rho_{12}\rho_{23}\cos(\varphi_{23} - \varphi_{12} + 2n_2\eta)}{\exp(2k_2\eta) + \rho_{12}^2\rho_{23}^2 \exp(-2k_2\eta) + 2\rho_{12}\rho_{23}\cos(\varphi_{23} + \varphi_{12} + 2n_2\eta)}$$

$$T = \frac{n_3}{n_1} \frac{\tau_{12}^2 \tau_{23}^2}{\exp(2k_2\eta) + \rho_{12}^2\rho_{23}^2 \exp(-2k_2\eta) + 2\rho_{12}\rho_{23}\cos(\varphi_{23} + \varphi_{12} + 2n_2\eta)}$$

(1)

where these parameters are in the form of:

$$\rho_{12}^2 = \frac{(n_1 - n_2)^2 + k_2^2}{(n_1 + n_2)^2 + k_2^2}, \rho_{23}^2 = \frac{(n_3 - n_2)^2 + k_2^2}{(n_3 + n_2)^2 + k_2^2}, \varphi_{12} = \arctan(\frac{2n_1k_2}{n_2^2 + k_2^2 - n_1^2})$$

$$\varphi_{23} = \arctan(\frac{2n_3k_2}{n_2^2 + k_2^2 - n_3^2}), \tau_{12}^2 = \frac{4n_1^2}{(n_1 + n_2)^2 + k_2^2}, \tau_{23}^2 = \frac{4(n_2^2 + k_2^2)}{(n_3 + n_2)^2 + k_2^2}$$

(2)

with (n_i, k_i) being refractive index and attenuation index of material i, and $h = (\lambda_0 / 2\pi)\eta$ being the thickness of metal layer.

In order to estimate the value of resistivity for ultra-thin metal layer, the reduction of electron mean-free-path (MFP) from that of bulk metals must be considered. Since the thickness of metal film (and thus also generally the grain size) is on the scale of tens of nanometers, the impact of excess scattering is very important. Since $\rho = \rho_{bulk} + \rho_{surface}$ with $1/\rho_{bulk} = A_1 *$ MFP and $1/\rho_{surface} = A_2 *$ MFP where we assume $A_1 = \beta A_2$, we get:

$$\rho = \rho_{bulk} * (1 + \beta\frac{MFP}{thickness})$$

(3)

Maximum anode spacing

From the electrical device perspective, the conductivity of transparent anode layer affects the carrier transport rates dramatically. Consider a device which contains anode contact (with the length of a) and total width of L, for small forward bias, the voltage drop across the device can be expressed as:

$$I_{lat}(y) = \int_0^y I_{ph}dy' = y * I_{ph} \rightarrow \Delta V(y) = -\int_{\frac{L-a}{2}}^y I_{lat}(y')\rho dy' = \frac{\rho I_{ph}}{2}[(\frac{L-a}{2})^2 - y^2] \quad (4)$$

Since we want the voltage drop across the device to be no larger than thermal voltage for minimal degradation, we get the criteria for device performance:

$$\Delta V_{max} \leq \frac{kT}{q} \rightarrow \frac{\rho I_{ph}}{8}(L - a)^2 \leq \frac{kT}{q}$$

(5)

Considering the fact that $\rho_{\square, metal} \ll \rho_{\square, PEDOT}$, we finally get:

$$L = 2\sqrt{\frac{2kT}{q}\frac{1}{\rho I_{ph}}} + a \approx 2\sqrt{\frac{2kT}{q}\frac{1}{\rho_{\square,metal}I_{ph}}} + a$$

(6)

Device simulation

By defining electric field $\vec{E}$ to be a gradient field of a scalar potential field Φ: $\vec{E} = -\nabla\Phi$, the Maxwell equation $\nabla \cdot \vec{E} = \frac{\rho}{\varepsilon}$ can be expressed as Poisson equation coupled with trap states:

$$\nabla \cdot \varepsilon\nabla\Phi = -\rho = -q(p - n) - \rho_{trap} \qquad (7)$$

By applying the divergence operator ∇ to the Maxwell's equation $\nabla \times \vec{H} = \vec{J} + \frac{\partial\vec{D}}{\partial t}$, and noting that $\vec{J} = \vec{J_n} + \vec{J_p}$, we can get the electrons and holes continuity equations:

$$\begin{cases} \frac{\partial n}{\partial t} = \frac{1}{q}\nabla \cdot \vec{J_n} + G_n - R_n \\ \frac{\partial p}{\partial t} = -\frac{1}{q}\nabla \cdot \vec{J_p} + G_p - R_p \end{cases} \qquad (8)$$

Finally, for organic materials, the electrons and holes excited by incident light are actually bound together, forming excitons [10, 11]. So we apply singlet-exciton continuity equation to represent the behavior of singlet excitons that can diffuse to P3HT / PCBM interface and dissociate into electrons and holes:

$$\frac{\partial N_{ex}}{\partial t} = \nabla \cdot D_{ex}\nabla N_{ex} + G_{ex} - R_{ex} - \frac{N_{ex} - N_{ex}^{equilibrium}}{\tau} \qquad (9)$$

where τ is the exciton lifetime, and D_{ex} is the exciton diffusion constant.

Sentaurus electrical device simulation includes Eq. (7), (8) and (9) to simulate the electrical behavior of organic solar cells, while solving for electric field and electrostatic potential distribution inside the device to give device performance such as short-circuit current (Jsc), open-circuit voltage (Voc) and fill-factor (FF).

DISCUSSION

Choice of transparent anode material

Table I. Optical properties of bulk metals for wavelength of 550 nm

Metal	n	k	MFP [nm]	Resistivity [$\Omega \cdot$nm]
Silver	0.2	3.44	52	15.9
Gold	0.47	2.83	39	22.1
Copper	0.62	2.57	38	16.8
Aluminum	1.44	5.23	14.9	26.5
ITO	1.92	0.048	27	1500

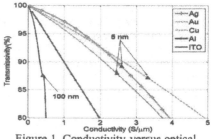

Figure 1. Conductivity versus optical transmissivity for different metals

Table I shows the basic parameters of several metals as bulk material. Comparing the experimental data that $\rho_{\square,silver} = 20\Omega/\square$ for silver thickness equals 10 nm [8], we get $\beta \approx 2$. Eq. (3) illustrates that the resistivity increases dramatically as the thickness goes below the mean-free-path. Clearly from the resistivity point of view, thick metal films can transport carriers more efficiently. For the transmission perspective, however, thinner metal films are more transparent, and thus block less light. From Fig. 1, for those metals for which mean free path is large, 5 nm thickness gives >85% transmission and very good conductivity at the same time. However, for metals with small value of mean free path, such as Aluminum, even 3nm thickness will cause

>20% reflection/absorption. Thus, for subsequent analyses, we use silver as the transparent anode material, and aluminum as the cathode material.

Optimization of active layer thickness

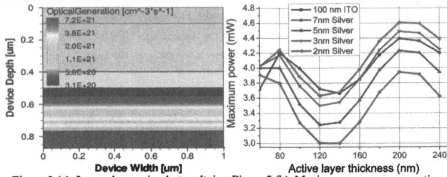

Figure 2 (a). Layered organic photovoltaic device structure with optical generation

Figure 2 (b). Maximum power versus active layer thickness at $\rho_{trap} = 10^{18} cm^{-3}$

Fig. 2 (a) shows the optical generation under stack structure, with materials being gas on top, glass as substrate, ultra-thin silver or 100 nm ITO as transparent anode layer, 50 nm PEDOT:PSS, P3HT:PCBM polymer as active layer and 100 nm aluminum as back metal layer. Here we assume normal incidence of sunlight with wavelength distribution following standard AM1.5G spectrum, energy bandgap of P3HT/PCBM being 1.5 eV [9], exciton diffusion length being ~10 nm [10, 11], exciton binding energy being 0.3 eV, and exciton lifetime being 100 ps [12]. From optical absorption point of view, the thicker the active layer is, the larger portion of incident light is absorbed. However, from the exciton recombination perspective, thicker active layer will cause larger portion of excitons and/or dissociated electrons and holes to recombine prior to collection by electrodes. Thus there is a trade-off in the active layer thickness. Fig. 2 (b) shows solar cell performance as a function of active layer thickness given a fixed value of trap states. Together with Fig. 2 (a), the best device performance corresponds to active layer thickness equaling 200 nm, in which there are two peaks of optical generation.

Device performance dependence on anode spacing and resistivity

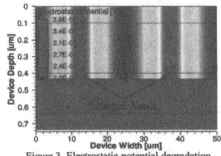

Figure 3. Electrostatic potential degradation of patterned anode structure

Table II. Metal thickness versus critical anode spacing in anode layer

Material	Thickness [nm]	Critical Anode Spacing [um]
ITO	50	8300
Silver	5	5400
Gold	5	5300
Copper	5	6100
PEDOT – PH500	36	25
Polymer with $\rho = 1\,\Omega \cdot cm$	50	100

Fig. 3 illustrates the electrostatic potential degradation around the anode due to the large resistivity of PEDOT. The anode width for a typical photovoltaic device is a = 5μm. From Eq. (6), by setting $I_{ph} = 10mA/cm^2$ for a standard organic solar cell, we get the critical spacing between neighboring anode metal grids, which is shown in Table II. For a silver anode layer with thickness equaling 5 nm, the critical anode spacing is 5400 μm. If we want the anode spacing to be larger than 100 μm and anode thickness to be 50 nm, the sheet resistance required for transparent polymers should be no larger than $2.3 * 10^5 \Omega/\square$, which is much larger than the sheet resistance of ITO, but smaller than that of PEDOT-PH500.

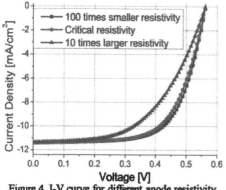

Figure 4. I-V curve for different anode resistivity

Figure 4 illustrates the device performance dependence on the anode layer resistivity. For 5 nm thick anode layer with 5 μm contact width and 50 μm device width, the critical resistivity calculated by Eq. (6) is $\rho_{critical} = 0.51\Omega\cdot cm$. As a comparison, we increase/decrease it by several times to show its impact on the I-V curves. It is clear that the device performance stays almost the same if the anode resistivity is lower than critical resistivity. However, both the maximum power and fill-factor decrease dramatically if the anode resistivity is higher than critical one. This shows that Eq. (6) serves as reasonable criteria for designing anode contact pattern.

Two-dimensional full device simulation

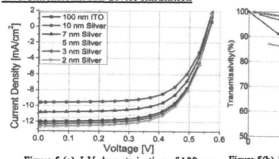

Figure 5 (a). I-V characterization of 100 nm ITO and different thickness of silver layer

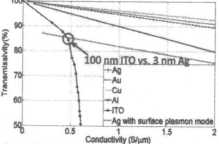

Figure 5(b). Transmissivity versus conductivity with / without surface plasmon mode turned on for metals

As another comparison, we use 100 nm ITO and different thickness of silver with electrical parameters calculated by Eq. (4) as the transparent anode layer. From the device simulation results in Fig. 5 (a), thinner silver layer gives better device performance, but only when the thickness reduces to less than 5 nm can we get the device performance to be equivalent to 100 nm ITO, which is consistent with the results in Fig. 2 (b), but contradicts Fig. 1 (b). The reason is that Eq. (1) is valid only under the assumption that materials in both sides are infinitely thick dielectric materials ($h_1 \rightarrow \infty$, $h_3 \rightarrow \infty$, $k_1 = k_3 = 0$). In solar cells, however, PEDOT has finite thickness and absorption coefficient. Also when the thickness of metal layer goes to nanometer range, the surface-plasmon effect is dramatic, which increases the absorption in the lossy metal layer. By turning on the surface plasmon mode, we get the reduced optical transmissivity shown in Fig. 5 (b), confirming that only 3 nm silver layer is equivalent to 100 nm ITO.

CONCLUSIONS

The relation between electrical resistivity and optical transmissivity of ultra-thin metals has been studied by considering the electron mean-free-path. By using Sentaurus Device simulator, we explored optimized thickness of active layer and its relation with the optical generation peaks inside active layer. Also, from the electrical device point of view, we calculated the critical resistivity of anode metal layer. The device simulation results show that silver layers thinner than 5 nm perform comparably to 100 nm ITO and suggest the possibility of replacing ITO with ultra-thin silver.

ACKNOWLEDGEMENT

The author would thank Jingyu Zou, Steven Hau, and Angus Yip in Professor Alex Jen's group in Department of Materials and Engineering, and Dr. Abhishek Kulkarni in Professor David Ginger's group in Chemistry Department in University of Washington, for their measurement of all the material parameters needed as the input of device simulation as well as the useful discussion during the collaboration.

REFERENCES

1. F. Padinger, R. S. Rittberger, and N. S. Sariciftci, *Adv. Funct. Mater.* 13, 85 (2003).
2. W. Ma, C. Yang, X. Gong, K. Lee, and A. J. Heeger, *Adv. Funct. Mater.* 15, 1617 (2006).
3. R. Kroon, M. Lenes, J. C. Hummelen, P. W. M. Blom, and B. Boer, *Polym. Rev.* 48, 531-582 (2008).
4. G. Li, V. Shrothiya, J. S. Huang, Y. Yao, T. Moriarty, K. Emery, and Y. Yang, *Nature Mater.* 4, 864 – 868 (2005).
5. D. Zhang, K. M. Ryu, X. L. Liu, E. Polikarpov, J. Ly, M. E. Tompson, and C. W. Zhou, *Nano. Lett.* 6, 1880 - 1886 (2006).
6. H. Y. Chen, J. H. Hou, S. Q. Zhang, Y. Y. Liang, G. W. Yang, Y. Yang, L. P. Yu, Y. Wu, and G. Li, *Nature Photon.* 3, 649 - 653 (2009).
7. M. Born, and E. Wolf, "An absorbing film on a transparent substrate," *Principles of Optics,* 7th edition, (Cambridge University Press) pp. 222 - 333.
8. V. J. Logeeswaran, P. K. Nobuhiko, P. Kobayashi, M. Saiflslam, W. Wu, P. Chaturvedi, N. X. Fang, S. Y. Wang, and R. S. Williams, *Nano Lett.,* 9, 178 – 182 (2009).
9. L. J. A. Koster, V. D. Mihailetchi, and P. W. M Blom, *Appl. Phys. Lett.* 88, 093511 (2006).
10. H. Hoppe, and N. S. Sariciftci, *J. Mater. Chem.* 16, 45 - 61 (2006).
11. S. R. Scully, P. B. Armstrong, C. Edder, J. J. J. Frechet, and M. D. McGehee, *Adv. Mater.* 19, 2961 - 2966 (2009).
12. J. Piris, T. E. Dykstra, A. A. Bakulin, P. H. M. Loosdrecht, W. Knulst, M. T. Trinh, J. M. Schins, and L. D. A. Siebbeles, *J. Phys. Chem. C.* 113, 14500 - 14506 (2009).
13. K. Yee, *IEEE T. Antenn. Propag.* 14, 302 – 307 (1966).
14. J. Berenger, *J. Comput. Phys.* 114, 185 – 200 (1994).
15. J. A. Roden, and S. D. Gedney, *Microw. Opt. Techn. Let.* 27, 334 – 339 (2000).

Oxide Interfaces

Mater. Res. Soc. Symp. Proc. Vol. 1270 © 2010 Materials Research Society   1270-HH15-05

Interfacial Dynamics of Perylene Derivatives attached to Metal Oxide Particle and Nanorod Films

Rainer Eichberger, Robert Schütz, Antje Neubauer, Thomas Hannappel and Andreas Bartelt
Helmholtz-Zentrum Berlin für Materialien und Energie, Solar Energy Research, Institute E-I5,
Hahn-Meitner-Platz 1, 14109 Berlin, Germany

ABSTRACT

Photo-induced electron transfer and recombination was investigated for prototype molecular absorbers covalently bound to metal-oxide films. Two perylene derivatives with different anchor/bridge groups, a propionic and an acrylic acid, were adsorbed on both bare ZnO nanorods and TiO$_2$ coated nanorods and the dynamics of the systems compared.

INTRODUCTION

Wide-gap metal oxide nanoarchitectures on the basis of ZnO and TiO$_2$ are subject to a rapidly growing number of scientific studies due to their economical importance as well as ecological attractiveness. In particular, low-cost concepts involving organic/metal oxide semiconductor photovoltaics [1] have moved into the focus of research as a result of increasing environmental attention. The best TiO$_2$ dye-sensitized solar cells (DSSC) to date exhibit efficiencies of 11% [2,3]. The dynamics of photo-induced heterogeneous electron transfer processes in dye-sensitized TiO$_2$ and ZnO has been studied extensively to gain fundamental insight into electron injection based solar cells [4-11]. It was recently reported that both metal-oxides show quite similar trends for the electron transfer reaction when sensitized by prototype perylene derivatives [12]. Electron transfer times range from 10 fs to 300 fs for these systems. ZnO has a band gap similar to that of TiO$_2$ and is considered to be a promising alternative material in DSSCs. Especially the absence of grain boundaries in crystalline ZnO nanorods [13-14] and a higher mobility [15-16] predict better directional transport properties in comparison with colloidal TiO$_2$. Also in combination with a solid hole conducting medium, nanorod morphology allows for better phase infiltration. Although efficiencies as high as 6,7 % have been reported for ZnO based DSSCs [17], the performance of TiO$_2$ solar cells remains superior. Many authors have reported on aggregate formation at ZnO electrodes due to removal of Zn^{2+} ions from the ZnO lattice during dye loading, which affects the heterogeneous electron transfer [18-21]. We have prepared both bare and TiO$_2$ dressed ZnO nanorods to address this problem.

In this work, we present time-resolved data obtained from fs transient absorption on sensitized bare and TiO$_2$ overcoated ZnO nanorod films in ultra-high vacuum (UHV) to exclude complications due to uncontrolled environmental parameters. Two prototype perylene derivatives were used as sensitizers to study the influence of the metal oxide semiconductors (Fig. 1). The decay of the donor excited state was measured simultaneously with the rise of the cationic state employing a white light probe continuum (WLC) spectrally covering both absorption bands that are energetically well separated, underlining the model character of these sensitizers (Fig. 2). The dynamics of the initial electron injection and early back-transfer is compared. The experiments show that the interfacial reaction dynamics for both bare ZnO and TiO$_2$ overcoated ZnO nanorods are very similar and comparable to earlier findings for films prepared from colloidal particles [12].

EXPERIMENT

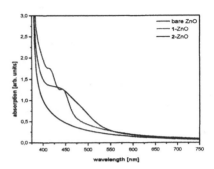

Figure 1. (a) Structures and abbreviations of the perylene derivatives **1** and **2**, (b) Ground state absorption spectra of dye sensitized and of bare ZnO nanorods

ZnO nanorods were deposited on a 40 nm thick ZnO seeding layer on an alkaline-free glass substrate. The nanorods were grown by chemical bath deposition in an aqueous solution of 0.01 M $Zn(NO_3)_2$ and 0.4 M NaOH at 80 °C [22]. The nanorods were rinsed in purified ethanol and annealed under ambient atmosphere at 450 °C. Their average length was about 1.6 μm, the mean diameter about 60 nm. For TiO_2-coating (Fig 4b), the ZnO nanorod electrodes were exposed to various cycles of spray pyrolysis of an isopropanolic solution of 0.02 M $Ti[OCH(CH_3)_2]_4$ and annealing at 450 °C. This procedure led to a closed coverage of the ZnO rods with an about 20 nm amorphous TiO_2 film. For sensitization the electrodes were heated in a muffle furnace at 430 °C in air for 45 minutes, cooled to about 50 °C under nitrogen atmosphere and immersed for 30 minutes in 2×10^{-4} M toluene solution of the dyes. The synthesis of the perylene derivatives 3-(8,11-di-*tert*-butyl-perylene-3-yl)propionic acid (perylene **1**) and 3-(8,11-di-*tert*-butyl-perylene-3-yl)acrylic acid (perylene **2**) is described elsewhere [23]. The dye coated films were rinsed with dried toluene for over 10 minutes. Transient absorption was performed using a 150 kHz regenerative amplifier providing 50 fs pulses of 800 nm. 85% of the output was used to generate sub 30-fs pulses at a central wavelength of $\lambda = 440$ nm using a near-infrared noncollinear optical parametric amplifier (NOPA) in combination with a second harmonic generation stage [24]. Pump pulse fluence was kept low and all measured signals were clearly in the linear regime. The remaining output from the regenerative amplifier was focused into a 2 mm sapphire crystal to generate a 450-1000 nm white light continuum probe (WLC). Time resolution was around 50 – 100 fs (fwhm) depending on the probe wavelengths [25-26]. Transient absorption was measured by lock-in with a photodiode using a monochromator for wavelength selection.

RESULTS AND DISCUSSION

Transient absorption spectra of 1-ZnO nanorods and 2-ZnO nanorods are shown in Fig. 2a and 2b, respectively. The black trace indicates that no absorption signal was present before the pump pulse fired. Isosbestic points between 600 nm and 625 nm indicate that the cation and excited state are only two active species. The spectral range from 630-750 nm shows the temporal development of the excited state as an instantaneous rise followed by a decay from 100 fs – 2 ps for both samples. The range 550-620 nm shows the rise of the cationic signal due to electron transfer from the excited state of the dyes into the semiconductor. The negative signal below 530 nm shows stimulated emission that decays as the electron is injected across the interface into ZnO.

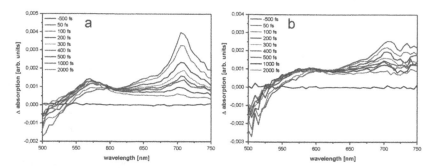

Figure 2. (a)Transient absorption spectra of 1-ZnO nanorods; (b) Transient absorption spectra of 2-ZnO nanorods. The pump wavelength was 440 nm for all spectra, the probe was the WLC.

Both nanorod samples 1-ZnO and 2-ZnO follow the same basic trend previously found for the identical molecules attached to mesoporous ZnO and TiO$_2$ films prepared from colloidal precursors [12]. The main spectral differences are related to **2**, where the specific nature of the bridge unit allows for delocalization of the excited state wavefunction inducing a stronger electronic coupling to the metal oxide semiconductor [27]. Such small chemical differences can be responsible for dramatic differences in electron transfer times at TiO$_2$ interfaces, but no distinct difference was found here for the injection times into ZnO nanorods. This is illustrated in Fig. 3a showing a transient measured at the maximum of the cation absorption. The rise times for the two molecules under study are virtually the same until about 1 ps when recombination sets in. A fit comprising a mono-exponential rise and a bi-exponential decay suggests an electron transfer time of around 350 fs. In comparison, on anatase TiO$_2$, bridge dependent electronic coupling is known to promote the electron from the excited donor into the semiconductor surface within 10 fs and 60 fs for molecule **1** and **2**, respectively [27]. Partial electronic state mixing between the organic and semiconductor component is responsible for the ultrafast electron transfer times. Injection times ranging from 200 fs to 300 fs for the same two perylene derivatives on ZnO nanoparticle films underline the electronic coupling effect also for the colloidal electrode [12]. For the nanorods however, the cation rise appears to be a little longer and the different electronic coupling effects of the specific bridge groups completely disappear (Fig. 3a).

The electron transfer process competes with a variety of excited state depopulation mechanisms that can lead to premature electron-hole recombination at the interface. In a working DSSC a surrounding electrolyte can alter electron transfer times by orders of magnitude by charge screening or interfacial electronic level realignment [28]. Also surface defects can hinder efficient electron escape into the bulk of the semiconductor. For cells with solid state hole collecting materials and lower screening potential, the critical time constants are expected to be much faster. Our measurements performed in UHV can be considered as a border situation since no hole conducting phase is added (the injected electron recombines with the geminate hole).

Fig. 3a also shows early electron-hole recombination monitored as the cation decay, which sets in after about 1 ps.

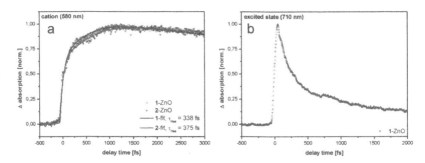

Figure 3. (a) Transient rise of the cationic state for 1-ZnO and 2-ZnO and the corresponding fits of the signals. The monoexponential rise give a time constants around 350 fs. (b) Transient decay of the excited state for 1-ZnO. The pump wavelength was 440 nm for both traces.

The ZnO nanorod surface seems to impose a limit onto the electron injection time: The coupling mechanism leading to ultrafast heterogeneous electron transfer observed for molecules 1 and 2 on TiO_2 and to some extent on ZnO colloidal films is completely washed out for the nanorods under study. Fig. 3b depicts the transient of 1-ZnO taken at the spectral maximum of the excited state absorption from Fig. 2a. The decay of this signal is a complementary indirect measurement of the electron injection process and corresponds the cation rise time.

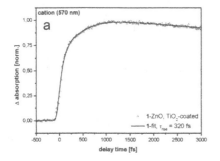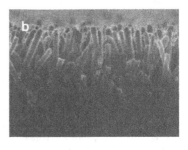

Figure 4. (a) Transient rise of the cationic state for TiO_2 overcoated 1-ZnO nanorods and the corresponding fit of the signal. The monoexponential rise gives a time constant of 320 fs. The pump wavelength was 440 nm (b) SEM picture od TiO_2 overcoated 1-ZnO nanorods.

The predominant surface for dye adsorption on the ZnO nanorods is the (10-10) face which has been widely studied with a variety of surface tools [29-30]. This surface can undergo severe structural and electronic changes when exposed to H atoms [31]. The surface near material behavior may vary from insulating to metallic depending on coverage. Adsorption of dyes with carboxylic acid groups from solution is assumed to proceed by forming a dissociative bond [12] while releasing hydrogen that can react with the surface. Indeed, many studies have reported on degradation of ZnO surfaces due to dye adsorption from solution, concluding that loading times must be minimized to avoid the formation of dye/Zn(II) aggregates that obstruct electron injection into the semiconductor [32]. Also, surface degradation can continue progressively within a DSSC containing an electrolyte. In our measurements performed in UHV, surface deterioration is kept to a minimum. To take advantage of both the higher conductivity of the crystalline ZnO core material [15] and the well known chemical stability of the TiO_2 surface we overcoated the ZnO nanorods with TiO_2. SEM images of the studied samples were taken to make sure that the nanorods were fully covered with a dense shell (Fig. 4b). Transient absorption data for this core-shell electrode sensitized with dye 1 is shown in Fig. 4. Surprisingly, on all observed time scales the dynamics are almost indistinguishable from the corresponding measurements on ZnO nanorods without the TiO_2 interlayer. Even shell thickness variations between a few and 20 nm did not bear any significant differences. XRD measurements taken on the samples after completion of the transient absorption measurements show that the TiO_2 coating mainly consisted of an amorphous phase. Since a high number of surface traps can be expected, most photo-excited electrons are presumably first injected into metal oxide semiconductor surface states prior to escape into the ZnO core or prior to the competing recombination reaction. For electrons that have escaped from the surface region it is assumed that the high mobility of the ZnO crystalline core structure can be advantageous for the final electron extraction in a DSSC. At present, it is not clear how fast electron transport from the surface to the bulk of these nanorods effectively is. For clean TiO_2 rutile surfaces sensitized with 1 or 2, ultrafast surface to bulk escape times well below 100 fs were reported from two-photon photoemission results [33]. Analog experiments (unpublished) performed on clean ZnO (10-10) single crystal surfaces indicate bulk escape on a similar time scale. A much rougher surface structure has to be assumed for the nanorods grown from a chemical bath and also for the ZnO architectures overcoated with TiO_2. Poorer electronic coupling to these surfaces is considered as one of the probable reasons for the observed longer electron injection times. Nevertheless, the absolute time scale for the electron transfer reaction is still very fast and encouraging for the further improvement of such core-shell photovoltaic interfaces. In a next approach efforts will be made to prepare a more crystalline TiO_2 coating to decrease the electron surface to bulk escape time.

CONCLUSIONS

We have presented fs transient absorption measurements on sensitized bare and TiO_2 overcoated ZnO nanorod films in UHV. Photo-induced electron transfer and recombination was investigated for perylene derivatives covalently bound to these metal-oxide films via a carboxylic anchor group. The dynamics of the systems was compared.

ACKNOWLEDGMENTS
The authors thank Ursula Michalczik and Sven Kubala for helpful assistance.
We are also grateful for financial support from the BMBF (Grant # 03SF0339G).

REFERENCES

1. K. Walzer, B. Maennig, M. Pfeiffer, K. Leo, *Chem. Rev.* **107**, 1233 (2007)
2. M. K. Nazeeruddin, F. De Angelis, S. Fantacci, A. Selloni, G. Viscardi, P. Liska, S. Ito, B. Takeru and M. Graetzel, *J. Am. Chem. Soc.*, **127**, 16835 (2005)
3. Y. Chiba, A. Islam, Y. Watanabe, R.Komiya, N. Koide and L. Han, *Jpn. J. Appl. Phys.* **45**, L638 (2006)
4. R. Ernstorfer, L. Gundlach, S. Felber, W. Storck, R. Eichberger, F. Willig, *J. Phys. Chem.* **B 110**, 25383 (2006)
5. R. Huber J. E. Moser, M. Graetzel, J. Wachtveitl, *J. Phys. Chem.* **B 106**, 6494 (2002)
6. Asbury, J.; Ellingson, R.; Ghosh, H.; Ferrere, S.; Nozik, A.; Lian, T. *J. Phys. Chem.* **B 103**, 3110 (1999)
7. J. Tornow, K. Schwarzburg, *J. Phys. Chem.* **C 111**, 8692 (2007)
8. J. Asbury, Y. Wang, T. Lian, *J. Phys. Chem.* **B 103**, 6643 (1999)
9. C. Bauer, G. Boschloo, E. Mukhtar, A. Hagfeldt, *J. Phys. Chem.* **B 105**, 5585 (2001)
10. R. Willis, C. Olson, B. O'Regan, T. Lutz, J. Nelson, J. Durrant, *J. Phys. Chem.* **B 106**, 7605 (2002)
11. A. Furube, R. Katoh, K. Hara, S. Murata, H. Arakawa, M. Tachiya, *J. Phys. Chem.* **B 107**, 4162 (2003)
12. J. M. Szarko, A. Neubauer, A. ABartelt, L. Socaciu-Siebert, F. Birkner, K. Schwarzburg, T. Hannappel, R. Eichberger, *J. Phys. Chem.* **C 112**, 10542 (2008)
13. J. B. Baxter, E. S. Aydil, *Appl. Phys. Lett.* **86**, 053114 (2005)
14. M. Law, L.E. Greene, J. C. Johnson, R. Saykally, P. Yang, *Nat. Mater.* **4**, 455 (2005)
15. E. Galoppini, J. Rochford, H. Chen, G. Saraf, Y. Lu, A. Hagfeldt and G. Boschloo, *JCP B*, **110**, 16159-16161 (2006)
16. J. B. Baxter, C. A. Schmuttenmaer, *J. Phys. Chem.* **B 110**, 25229 (2006)
17. B. Tan, Y. Y. Wu, *J. Phys. Chem.* **B 110**, 15932(2006)
18. K. Westermark, H. Rensmo, H. Siegbahn, K. Keis, A. Hagfeldt, L. Ojama and P. Persson, *J. Phys. Chem.* **B 106**, 10102 (2002)
19. K. Keis, J. Lindgren, S.-E. Lindquist, and A. Hagfeldt, *Langmuir,* **16**, 4688-4694 (2000)
20. H. Horiuchi, R. Katoh, K. Hara, M. Yanagida, S. Murata, H. Arakawa, and M. Tachiya, *J. Phys. Chem.* **B 107**, 2570 (2003)
21. T. P. Chou, Q. Zhang, G. Cao, *J. Phys. Chem.* **C 111**, 18804 (2007)
22. R. B. Peterson, C. L. Fields, B. A. Gregg, *Langmuir* **20**, 5114 (2004)
23. C. Zimmermann, F. Willig, S. Ramakrishna, B. Burfeindt, B. Pettinger, R. Eichberger, W. Storck, *J. Phys. Chem.* **B 105**, 9245 (2001)
24. J. Piel, E. Riedle, L. Gundlach, R. Ernstorfer, R. Eichberger, *Opt. Lett.* **31**, 1289 (2006)
25. A. Penzkofer, W. Falkenstein, *Opt. Commun.* **17**, 1 (1976)
26. P. Sathy, A. Penzkofer, *Appl. Phys. B: Laser Opt.* **1**, 127 (1995)
27. Ernstorfer, R., Ph.D. Thesis, Freie Universitaet Berlin, 2004
28. A. F. Bartelt, R. Schuetz, A. Neubauer, T. Hannappel, R. Eichberger, *J. Phys. Chem.* **C 113**, 21233 (2009)
29. O. Dulub, L. A. Boatner, and U. Diebold, *Surface Science* **519**, 201-217 (2002)
30. C. Woell, *Progress in Surface Science* **82**, 55 (2007)
31. W. A. Tisdale, M. Muntwiler, D. J. Norris, E. S. Aydil,X.-Y. Zhu, *J. Phys. Chem.* **C 112**, 14682 (2008)
32. T. P. Chou, Q. Zhang, G. Cao, *J. Phys. Chem.* **C 111**, 18804 (2007)
33. L. Gundlach, T. Letzig, F. Willig, J. Chem. Sci. 121, 561 (2009)

Devices I

Mater. Res. Soc. Symp. Proc. Vol. 1270 © 2010 Materials Research Society   1270-II01-08

Transient charge carrier transport effects in organic field effect transistor channels

Hsiu-Chuang Chang[1], P. Paul Ruden[1], Yan Liang[2], and C. Daniel Frisbie[2]
[1]Department of Electrical and Computer Engineering, University of Minnesota,
Minneapolis, MN 55455, U.S.A.
[2]Department of Chemical Engineering and Materials Science, University of Minnesota,
Minneapolis, MN 55455, U.S.A.

ABSTRACT

We present device simulations exploring the effects of traps during transient processes in the conducting channel of organic semiconductor field effect transistors (OFETs). The device structure considered resembles a typical organic thin-film transistor with one of the channel contacts removed. However, the channel length is much longer than in typical OFETs in order to increase the transit time. By measuring the displacement current in these long-channel capacitors, transient effects in the carrier transport in organic semiconductors may be studied. When carriers are injected into the device, a conducting channel is established while traps, which are initially empty, are being populated. The filling of the traps then modifies the transport characteristics of the injected charge carriers. In contrast, DC experiments as they are typically performed to characterize the transport properties of organic semiconductor channels investigate steady states with traps partially filled. Numerical and approximate analytical models for the formation of the conducting channel and the resulting displacement current are discussed. We show that displacement current measurements on OFET structures provide unique opportunities for the study of trap dynamics.

INTRODUCTION

Organic semiconductors are among the most promising materials for novel electronics. Many electronic devices have been fabricated from conjugated organic hydrocarbons, and some are already commercialized [1]. For carrier transport in organic materials, the trapping/detrapping of mobile charge carriers is an important issue that can even play the dominant role in device performance. Frequently, organic field effect transistors (OFETs) are used to characterize the electrical properties of organic semiconductors (OSCs) [2,3]. Usually one measures the DC current through the conducting channel as a function of the applied gate-to-source and drain-to-source voltages and obtains a steady state charge carrier field effect mobility. The steady state mobility is measured with traps partially filled and the dynamic effects of traps do not manifest themselves directly in the experiments. Transient measurements of the displacement current [4,5,6] can provide information about the dynamics of carrier trapping phenomena. An effective mobility that characterizes transient experiments is measured when trapping effects are in process and the fractional trap filling changes. In the displacement current measurements, we use a two-terminal device structure resembling a simplified field effect transistor with one of the channel contacts removed, as shown in Figs. 1(a) and 1(b). Under suitable bias a conducting channel is established in the device. Since the channel length is on the order of millimeters, charge carriers take a relatively long time to reach the far end of the device after having been injected by the channel contact. Transient effects that occur on a timescale comparable to the transit time may readily be explored. Trapping/detrapping processes take place once the carriers are injected into the OSC layer. During channel formation traps with different time constants have different effects on the effective carrier mobility. Slow traps do not change the mobility on the time scale of the experiment; fast traps decrease the effective mobility relative to the value that would be measured in the absence of traps. Most interesting is the intermediate case of traps with time constants comparable to the transit time. In that case, the effective mobility of the charge carriers varies with time.

DEVICE STRUCTURE

A schematic diagram of the device geometry is shown in Figs. 1(a) and 1(b). For all results discussed here the width of the pentacene strip (W) is 0.5 mm and the channel length (L) is 6 mm. Electrical characterization is carried out by measuring the displacement current running through the grounded top electrode while linearly sweeping the bottom (gate) electrode voltage ($V_B(t)$). $V_B(t)$ is taken to follow a simple zigzag time profile with sweep rates of $\mp 5.7\,\mathrm{Vs^{-1}}$. As shown in Fig. 1(c), the gate voltage decreases first and holes are injected into the device beginning at $t = 0$. For the discussion in this report (numerical simulations), we plot the displacement current into the top contact versus time to explore the transient effects. After the displacement current becomes constant, the sweep rate changes to a positive value (in Fig. 1(c) at $t = 4$ s), the gate voltage starts to increase, and holes are extracted out of the device. The extraction persists until the channel is essentially depleted at $t = t_d$. A detailed description of the experimental $I(t)$ vs. $V(t)$ characteristics can be found in a previous report [7].

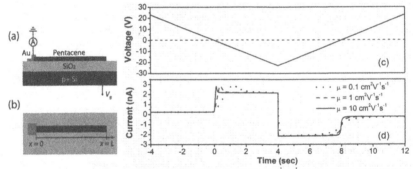

Figure 1. (a) Schematic cross section of the device structure (300 nm thick thermal oxide layer and 30 nm thick pentacene film). (b) Schematic top view of the device. (c) Applied gate voltage versus time. (d) Calculated I versus t characteristics using different hole mobilities in the full numerical model of a trap free case.

DEVICE MODEL

As shown in Fig. 1(b), the x-axis is parallel to the pentacene strip with its origin at the edge of channel contact. The gradual channel approximation is applicable because the device's lateral dimensions are large [8]. The local charge in the channel is proportional to the local voltage drop across the SiO_2, $V(x,t)$, and the lateral electric field in the channel is $-dV(x,t)/dx$. $V(x,t)$ is proportional to the sum of the sheet concentration of mobile holes $p(x,t)$ and trapped holes $p_T(x,t)$. The linear current density (J) is comprised of drift and diffusion current contributions. As holes are injected from the channel contact into the pentacene, $p(x,t)$ varies with time according to the continuity equation,

$$\frac{\partial p}{\partial t} = -\frac{1}{e}\frac{\partial}{\partial x}J + \frac{\partial p}{\partial t}\Big|_{traps} \tag{1}$$

The sheet concentration of trapped holes is controlled by the total (sheet) density of traps, N_T, the capture coefficient, $\tilde{c}$, and the emission coefficient, $\tilde{e}$ [9].

$$\frac{\partial p_T}{\partial t} = \tilde{c}p(N_T - p_T) - \tilde{e}p_T = -\frac{\partial p}{\partial t}\Big|_{traps} \tag{2}$$

We define the capture (τ_{cap}) and emission (τ_{emi}) time constants as the inverse of $\tilde{c}N_T$ and the inverse of $\tilde{e}$, respectively. τ_{cap} determines how fast free holes can be trapped and τ_{emi} determines how quickly trapped holes are released. To solve Eqs. (1) and (2) appropriate initial and boundary conditions are needed [7]. When bias is applied and holes are injected into the channel, the following condition at the contact holds for $V(0,t) > 0$:
$e[p(0,t) + p_T(0,t)] = C'V(0,t)$, where $V(0,t) = -V_B(t) = -r_V t$, r_V is the sweep rate, e is the elementary charge, C' is the SiO_2 capacitance per unit area, and $t > 0$.

RESULTS AND DISCUSSION

The coupled partial differential equations (1) and (2) need to be solved numerically. However, it is helpful to explore the case of negligible trap concentration first, i.e. $p_T \equiv 0$. Figure 1(d) shows simulation results with different mobilities under this condition. Evidently, higher mobilities yield shorter transients, i.e. time intervals over which the current varies. The hole injection continues for four seconds, then the voltage ramp rate reversal leads to carrier extraction, which again persists for about four seconds.

One may understand the transient displacement current versus time characteristics qualitatively on the basis of an analytical geometric model. This model considers only the drift current and completely neglects traps: At time t the hole concentration at $x = 0$ is $p(0,t) = (C'/e)|r_V|t$, and $p(x,t)$ is assumed to decrease linearly to zero at $x = x_0(t) = \mu Et$. The hole density profile is thus a triangle and the total hole concentration injected at time t is $P(t) = x_0(t)p(0,t)/2$. Since the lineal current density is $J(0,t) = e(d/dt)P(t)$, the slope of $I(t)$ in the transient region is proportional to the square root of the mobility. The transient regime extends over a time interval of $\Delta t = 2L/\sqrt{\mu|r_V|}$ in this simple model. To justify the focus on the drift current contribution, we display in Table I the relevant timescales for three different mobility values. Evidently, diffusion is too slow a process to be relevant for the voltage ramp rates and other device parameters adopted here.

113

Table I. Time scales for different carrier transport effects in transient process.

	Δt	$L^2/(\mu\, r_v \Delta t)$	$L^2/(kT\mu/e)$
$\mu = 0.1\ \text{cm}^2\text{V}^{-1}\text{s}^{-1}$	1.585 s	1.256 s	144 s
$\mu = 1\ \text{cm}^2\text{V}^{-1}\text{s}^{-1}$	0.5 s	0.126 s	14.4 s
$\mu = 10\ \text{cm}^2\text{V}^{-1}\text{s}^{-1}$	0.158 s	0.0126 s	1.44 s

If we plot the displacement current versus time according to the analytical geometric model, the transient current peaks are in reasonable agreement with the numerical results of Fig. 1(d). Furthermore, including the effects of traps will modify the picture, but we may still define an effective mobility (μ_{eff}) as given by the slope of $I(t)$ the transient region. It is useful to consider separately three different cases, according to the timescales characterizing the traps. In all subsequent simulations $\mu = 1.0\ \text{cm}^2/\text{Vs}$ is used.

Fast traps:

$\tau_{cap}, \tau_{emi} \ll \Delta t$. In this case rapid trapping and detrapping of the injected charge carriers leads to an average mobility, which for the case of non-degenerate statistics is given by:

$$\mu_{ave} = \mu\frac{\tau_{cap}}{\tau_{cap}+\tau_{emi}} \qquad (3)$$

Extraction of the effective mobility from the I(t) slope in the transient region then yields: $\mu_{eff} = \mu_{ave}$. Simulation results for this case are presented in Figure 2.

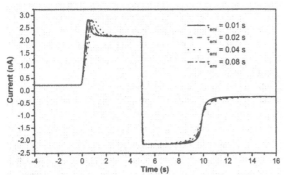

Figure 2. Calculated I versus t characteristics for different trap time constant that are much smaller than Δt: $\tau_{cap} = 0.01$ s, and $\tau_{emi} = 0.01, 0.02, 0.04,$ and 0.08 s.

Slow traps:

$\tau_{cap}, \tau_{emi} >> \Delta t$, $\mu_{eff} \cong \mu$. When the average trapping time is much larger than Δt, almost no mobile holes are captured during the transient and the mobility is unaffected by the traps. The transient region is therefore essentially unaffected, i.e. $\mu_{eff} \approx \mu$, but if the capture times are comparable or shorter than t_d shifts in the threshold where full depletion is achieved will appear in the $I(t)$ results. This is shown in Fig. 3.

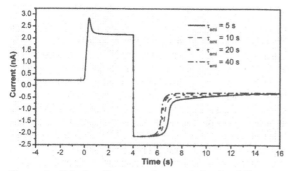

Figure 3. Calculate I versus t characteristics for different trap time constants that are much larger than Δt: $\tau_{cap} = 5$ s, and $\tau_{emi} = 5, 10, 20$ and 40 s.

Intermediate traps:

$(\tau_{cap} + \tau_{emi}) \approx \Delta t$, $\mu_{eff} = \mu_{eff}(t)$. When the average trapping time is comparable to the time scale of the current transient, the effective mobility is time dependent as the fraction of charge carriers trapped changes. In Fig. 4, calculated $I(t)$ characteristics for different trapping

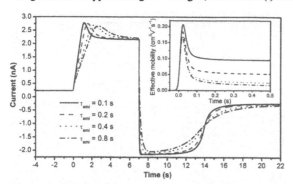

Figure 4. Calculated I versus t characteristics for trap emission time constants comparable to Δt: $\tau_{emi} = 0.1, 0.2, 0.4,$ and 0.8 s; $\tau_{cap} = 0.01$ s. Inset: μ_{eff} vs. time, for the cases discussed. Solid, dashed, dotted, and dash-dot lines are for $\tau_{emi} = 0.1, 0.2, 0.4,$ and 0.8 s, respectively.

115

parameters are plotted. The capture time is chosen to short (τ_{cap} = 0.01 s), but the emission time constant is varied, τ_{emi} = 0.1, 0.2, 0.4, and 0.8 s, and comparable to Δt. The effective mobility versus time is plotted in the inset of Fig. 4. Evidently, the effective mobility is larger at the beginning of charge injection and decreases subsequently. For the shorter emission times, near steady state conditions are approached towards the end of the time range plotted in the inset.

CONCLUSIONS

We discuss the effect of traps on the transient $I(t)$ characteristics of organic semiconductor field effect transistor structures. A simple geometric model of the injection process is used as a basis for the definition of an effective mobility. Because of charge carrier trapping, this effective mobility may be different from the free carrier mobility. It is also clearly different from the time-average mobility that can be extracted from steady state OFET data. Different time scales of carrier transport effects are compared. The wide range of time constants associated with carrier capture into traps and emission from traps allows for several distinct cases to be examined. Very fast traps (on the time scale of the transient region) give rise to an average mobility that may be obtained from either steady state or transient measurements of the kind discussed in this work. Trapping processes that are slow on the relevant timescale of the experiment have negligible impact on the effective mobility, but may manifest themselves as threshold shifts in the carrier extraction part of the $I(t)$ results. Most interesting is the case of traps with time constants that are comparable to the experimental transient time scale. Under these conditions the effective mobility characterizing the injection transient is itself time dependent.

ACKNOWLEDGMENT

This work was supported primarily by the MRSEC Program of the National Science Foundation under Award Number DMR-0212302 and DMR-0819885. Access to the facilities of the Minnesota Supercomputing Institute for Advanced Computational Research is gratefully acknowledged.

REFERENCES

1. H. E. Katz and J. Huang, Annu. Rev. Mater. Res. **39**, 71-92 (2009).
2. C. R. Newman, C. D. Frisbie, D. A. da Silva Filho, J.-L. Bredas, P. C. Ewbank, and K. R. Mann, Chem. Mater. **16**, 4436 (2004).
3. H. Sirringhaus, Adv. Mater. (Weinheim, Ger.) **17**, 2411 (2005).
4. S. Ogawa, Y. Kimura, H. Ishii, and M. Niwano, Jpn. J. Appl. Phys. **42**, L1275 (2003).
5. S. Suzuki, Y. Yasutake, and Y. Majima, Jpn. J. Appl. Phys. **47**, 3167 (2008).
6. E. Pavlica and G. Bratina, J. Phys. D: Appl. Phys. 41, 135109 (2008).
7. Y. Liang, C. D. Frisbie, H.-C. Chang, and P. P. Ruden, J. Appl. Phys., **105**, 024514 (2009).
8. S. M. Sze, *Physics of Semiconductor Devices*, 2nd ed., Wiley, New York (1981).
9. G. Horowitz, R. Hajlaoui, P. Delannoy, J. Phys. III France 5, 355 (1995).

116

Poster Session: Charge Transport and Devices

Mater. Res. Soc. Symp. Proc. Vol. 1270 © 2010 Materials Research Society   1270-II06-20

Hall effect of solution-crystallized and vapor-deposited
2,7-dioctylbenzothieno[3,2-b]benzothiophene field-effect transistors

M. Yamagishi[1], T. Uemura[1,2], Y. Takatsuki[1], J. Soeda[1], Y. Okada[1], Y. Hirose[1], Y. Nakazawa[1],
S. Shinamura[3], K. Takimiya[3], and J. Takeya[1,2]
[1]Graduate School of Science, Osaka University, Toyonaka 560-0043, Japan.
[2]ISIR, Osaka University, Ibaraki 567-0047, Japan.
[3]Graduate School of Engineering, Hiroshima University, Higashi-Hiroshima 739-8527, Japan.

ABSTRACT

Gate-voltage dependent Hall coefficient R_H is measured in high-mobility field-effect transistors of solution-crystallized and vapor-deposited 2,7-dioctylbenzothieno[3,2-b]benzothiophene. The value of R_H evolves with density of accumulated charge Q, precisely satisfying the free-electron formula $R_H = 1/Q$ near room temperature. The result indicates that the intrinsic charge transport inside the grains is band-like in the high-mobility organic-semiconductor thin films that are of significant interest in industry. At lower temperatures, even Hall-effect mobility averaged over the whole polycrystalline film decreases due to the presence of carrier-trapping levels at the grain boundaries, while the free-electron-like transport is preserved in the grains. With the separated description of the inter- and intra-grain charge transport, it is demonstrated that the reduction of mobility with decreasing temperature often shown in organic thin-film transistors does not necessarily mean mere hopping transport.

INTRODUCTION

Organic field-effect transistors (OFETs) have been intensively investigated because of their capability of applications to flexible, light-weight, large-area and low-cost next-generation electric devices. In terms of the maximum field-effect mobility, development of organic single-crystal transistors disclosed that even the value of more than 20 cm^2/Vs is reachable for practical OFETs and that the high-mobility transport is based on a diffusive transport. However, they have difficulties in producing commercial components because crystals are grown independently of substrates. Vapour-deposited polycrystalline thin films usually show one-order less performance, though they have moderate accessibility to large-scale application. On the other hand, solution processes are the most suitable to large-scale and low-cost fabrication but have suffered from low mobility. However, very recently, we develop a method to fabricate crystalline thin films from solution showing high carrier mobility exceeding a few cm^2/Vs. In order to understand carrier transport mechanism of practically useful solution-processed crystalline films and vapour-deposited polycrystalline films, we measured Hall effect, which can directly examine that the microscopic in-grain transport mechanism separated from effects of grain boundaries.

In the present experiments, we focus on recently synthesized semiconducting material, 2,7-dioctylbenzothieno[3,2-b]benzothiophene (C8-BTBT), which shows carrier mobilities as high as ~3 cm^2/Vs in both vapour-deposited polycrystalline films [1] and in solution-processed crystal films [2]. To construct OFETs for Hall effect measurement, Hall bars were shaped in the films fabricated on SiO$_2$ gate dielectrics. Results on both films show that inverse of Hall coefficient $1/R_H$ (= $I_D \Delta B / \Delta V_H$ = ne) well corresponds to the charge amount calculated from the simple capacitance model. The observation means that diffusive band transport is realized equally in the solution-crystallized films and inside grains of the vapor-deposited polycrystalline films. Therefore, relatively high ordering is achieved even in thermally deposited films inside the grains which lead to high field-effect mobility when grain boundaries do not affect seriously to restrict the carrier transport.

EXPERIMENT

The solution-crystallized C8-BTBT thin film is prepared using recently developed technique of the orientational growth [2]. A 0.4wt% solution of C8-BTBT is prepared with a solvent of heptane and a droplet is sustained at an edge of a structure on an inclined substrate, so that the crystalline domain grows in the direction of the inclination through evaporation of the solvent. Typical thickness of the crystal is 100 nm. In order to thoroughly remove the solvent, we dried it in a vacuum oven for typically 5 h at 50 C. To prepare the sample for the Hall-effect measurement, gold is deposited to form top-contact electrodes of the source, drain and additional pads to probe potentials in the channel. A laser-etching technique is employed for all the three samples to shape the channels for the Hall-bar configuration. Figure 1 shows the top view of the solution-crystallized C8-BTBT thin-film device (Sample A) for the measurement. The channel length L is about 240 μm and the width W is about 60 μm. The distance between the two voltage-probing points for the four-terminal conductivity measurement is approximately 120 μm. Thus, conductivity σ is estimated as

$\sigma = I_D / (V_2 - V_1)$, using measured drain current I_D and the voltages at the two points V_2 and V_1. With the application of magnetic field B perpendicular to the channel, we measure the voltage V_3 on the other side to estimate R_H. Detecting modulation ΔV_H of $V_3 - V_2$ with that of B, Hall coefficient is defined as $\Delta V_H / I_D \Delta B$.

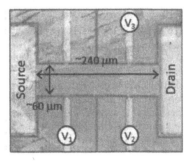

Figure 1. Top view of the solution-crystallized C8-BTBT thin-film device (Sample A) for the Hall-effect measurement.

Figure 2 shows surface topography of the crystalline thin film measured by atomic-force microscope (AFM). Molecularly flat region extends over more than a few μm, indicating the high crystallinity of the film. Transfer and output characteristics are shown in Figure 3, where well-defined transistor performance is demonstrated. The mobility estimated from the slope of current regression with the gate voltage V_G, so that the value reached as high as 6 cm^2/Vs. We note that the value is remarkably high against the standard of solution-prepared thin-film devices and has been only achieved with vapor-crystallized single-crystal devices.

Figure 2. Atomic-force microscope view of the solution-crystallized C8-BTBT thin-film transistor.

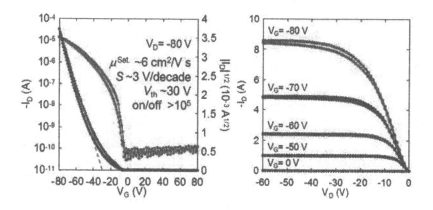

Figure 3. Transfer and output characteristics of a solution-crystallized C8-BTBT transistor.

121

Vapor-deposited sample (Sample B) of C8-BTBT is prepared to the same shape and size as the solution crystallized Sample A. The thin film is deposited to the thickness of approximately 50 nm with the rate of approximately 0.05 nm/min. Similarly, laser-etching technique is employed to shape the sample to the Hall bar. The AFM view is shown in Figure 4, where relatively good connectivity between grains is visible. Therefore, mobility is as high as 3 cm^2/Vs.

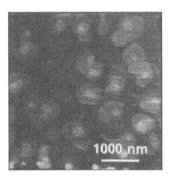

Figure 4. Atomic-force microscope view of the vapor-deposited C8-BTBT thin-film transistor.

RESULTS AND DISCUSSION

Hall effect of the solution-crystallized C8-BTBT thin-film transistor

Figures 5 shows the result of simultaneous measurements of the Hall effect and the four-terminal conductivity σ in Sample A. The plot of the inverse Hall coefficient as a function of gate voltage at room temperature basically follows that of σ. $1/R_H$ increases with negative V_G in accordance with the free-electron formula $1/R_H = CV_G$ within the accuracy of the measurement, presenting a clear indication of the diffusive charge transport. C is the capacitance of the gate-dielectric insulator SiO_2. The four-terminal sheet conductivity monotonically evolves with negative V_G due to the hole accumulation. We define field-effect mobility μ_{FET} on the four-terminal measurement $1/C \ d\sigma/dV_G$ and Hall-effect mobility μ_H as $R_H\sigma$. The single-crystal sample satisfies the free-electron formula in the Hall-effect measurement, so that μ_{FET} is identical to μ_H. The whole results resemble those for high-mobility rubrene single crystals with $\mu_{FET} \sim 8$ cm^2/Vs at room temperature, which we reported previously [3].

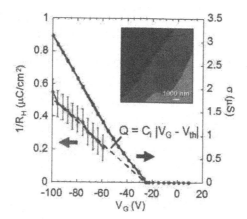

Figure 5. Inverse Hall coefficient and conductivity as a function of gate voltage for the solution-crystallized C8-BTBT transistor.

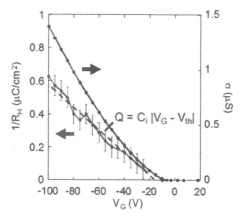

Figure 6. Inverse Hall coefficient and conductivity as a function of gate voltage for the vapor-deposited C8-BTBT transistor.

Hall effect of the vapor-deposited C8-BTBT thin-film transistor

Figure 6 presents the plot for the device with the slowly deposited polycrystalline thin film (Sample B). Similarly as the result of the single-crystal device, $1/R_H$ increases with negative V_G following the slope defined by C. The result indicates that all the mobile charge is again diffusive with spatially extended electronic states. Due to the microscopic origin of the Hall effect, $1/R_H$ is not strongly influenced by the presence of grain boundaries as long as their inter-grain electrical connection is fairly good, which is demonstrated in the present result itself.

Finally temperature dependence of the mobilities for Samples A and B are presented in Figure 7. Though the fundamental transport mechanism is diffusive band transport as evidenced in the Hall effect, the temperature dependence significantly differs between those for the two samples. The result indicates that grain-boundary traps are responsible in the decreasing mobility with decreasing temperature in Sample B. Therefore, such temperature dependence does not mean mere hopping transport in this case.

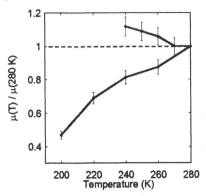

Figure 7. Temperature dependence of mobility in solution-crystallized and vapor-deposited C8-BTBT thin-film transistors.

CONCLUSIONS

Simultaneous measurement of the Hall coefficient and four-terminal conductivity in the single- and poly-crystalline samples of the C8-BTBT field-effect transistors has provided separated description on the intra- and inter-grain charge transport. The electronic states are basically extended over molecules so that the band-like diffusive transport is realized, exhibiting the free-electron Hall effect. The above feature is qualitatively same in a single domain of the poly-crystalline film deposited on a substrate, though the traps in grain boundaries become increasingly significant in averaged value of mobility at low temperatures.

ACKNOWLEDGMENTS

DNTT is supplied from Nihon Kayaku Co., ltd.. This work was financially supported by Industrial Technology Research Grant Program from NEDO, Japan, and by a Grant-in-Aid for Scientific Research (Nos. 21108514 and 22245032) from MEXT, Japan.

REFERENCES

1. T. Izawa, E. Miyazaki, and K. Takimiya, *Adv. Mater.* **20**, 3388 (2008).
2. T. Uemura, Y. Hirose, M. Uno, K. Takimiya, and J. Takeya, *Appl. Phys. Exp.* **2**, 111501(2009).
3. J. Takeya, J. Kato, K. Hara, M. Yamagishi, R. Hirahara, K. Yamada, Y. Nakazawa, S. Ikehata, K. Tsukagoshi, Y. Aoyagi, T. Takenobu, and Y. Iwasa, *Phys. Rev.Lett.* **98**, 196804 (2007).

Mater. Res. Soc. Symp. Proc. Vol. 1270 © 2010 Materials Research Society　　　　1270-II06-46

Titanyl Phthalocyanine (TiOPc) Organic Thin Film Transistors with Highly π-π Interaction

Huang-Ming P. Chen, Yung-Hsing Chen and Bo-Ruei Lin
Department of Photonics & Display Institute, National Chiao Tung University, Hsinchu, Taiwan.

ABSTRACT

The objective of this research is to obtain uniform vacuum-deposition triclinic phase II crystal of titanyl phthalocyanine (α-TiOPc) films from various TiOPc crystal forms. The crystal structure and morphology of vacuum-deposited TiOPc films can be manipulated by deposition rate and substrate temperature. Crystal structure was determined by X-ray diffraction (XRD). Thin film morphology was analyzed by scanning electron microscope (SEM). Highly ordered α-TiOPc film with an edge-on molecular orientation was deposited on octadecyltrichlorosilane (OTS) treated Si/SiO₂ surface. All TiOPc crystal forms, such as amorphous, α and γ phases, provided the triclinic phase II crystal of TiOPc. The full width at half maximum (FWHM) of the peak at 7.5 degree in XRD spectra was 0.23, 0.27 and 0.29 for γ, α and amorphous powder with substrate temperature maintained at 180°C, respectively. The FWHM of the 7.5 degree peak can be achieved 0.22 deposited from all crystal forms at elevated temperature higher than 220°C. The α-TiOPc deposition film exhibited an excellent p-type semiconducting behavior in air with dense packing structure due to the close π–π molecular packing. The devices, field-effect mobility range from 0.02 to 0.26 cm²/V s depending on various process parameters. The on/off current ratio (I_{on}/I_{off}) is over 10^5. The TiOPc OTFTs will be applied as multi-parameter gas sensor in the near future.

INTRODUCTION

Organic thin film transistors (OTFTs) have experienced a rapid development in the past decade. Their potential applications include electronic paper, flexible display, sensors, low-end smart card and radio frequency identification tags (RFIDs) [1]. Organic materials possess excellent advantages over inorganic materials, such as low processing temperature, large area coverage, structural flexibility, good compatibility with different substrates, and low costs. Among various organic semiconductors, phthalocyanine compounds have attracted considerable attentions in OTFTs owing to their remarkable chemical and thermal stabilities, non-toxicity and good field-effect performance [2, 3]. Since the first OTFT was reported in 1986 [4], significant attention have been paid to the phthalocyanine-based OTFTs. Pc-based OTFT was first reported on scandium diphthalocyanine (ScPc₂) in 1988 [5]. The mobility of ScPc₂ was, however, only 10^{-3} cm²/V s. Titanyl phthalocyanine (TiOPc), a polar, non-planar, and pyramid-like molecule, exhibited an n-type semiconducting behavior in ultra high vacuum and a p-type semiconducting behavior under air condition [6]. It is due to its close π–π molecular packing [7]. Tree common crystal structures of TiOPc are monoclinic phase I (β -TiOPc), triclinic phase II (α -TiOPc), and triclinic phase Y (γ -TiOPc) [8]. According to the literature, the triclinic α -structure of TiOPc has significant molecular overlaps and possesses very short intermolecular distances. In this study, we focused on finding the optimized process conditions to obtain thin α-TiOPc film from different crystal forms, which may beneficial to the future TiOPc TFTs device applications.

EXPERIMENT

TiOPc OTFTs with bottom-gate top-contact were fabricated according to the following procedure. Heavily doped, n-type Si (100) wafer was used as gate electrode and substrate. On the surface of silicon, the SiO_2 layer with the thickness of 100 nm was served as gate insulator by thermal growth. The Si/SiO_2 wafer was ultrasonically cleaned with acetone, detergent and deionized water, respectively. Then the clean Si/SiO_2 wafer was dried at 100°C for 1 h in order to eliminate the influence of the moisture. Treatment of Si/SiO_2 wafer with octadecyltrichlorosilane (OTS) was prepared by vapor deposition method at 120°C for 2 h. TiOPc film was deposited on the substrate by thermal evaporation. Three crystal forms of TiOPc such as amorphous, α and γ phases were used, respectively. All crystal forms of TiOPc were not further purified. The deposition rate was fixed at 0.1~0.3 Å/s and the thickness of TiOPc film was 60 nm. For this TiOPc evaporation process, the substrate deposition temperatures between 150°C and 220°C. The pressure of vacuum chamber in the evaporation system was 3×10^{-6} Torr. Subsequently, 50 nm thick gold source/drain electrodes were deposited on TiOPc film via a shadow mask. The channel length and width were 0.1 mm and 2.0 mm, respectively.

DISCUSSION

X-ray diffraction patterns of TiOPc films deposited on OTS treated Si/SiO_2 surfaces at 180°C from various TiOPc crystal forms, such as amorphous, α - and γ-TiOPc, are shown in Figure 1a. There are an intense peak and a weak peak at 7.5° and 15° in XRD spectra due to the diffraction from (010) and (020) plane of α phase. The XRD spectra indicated that all different crystal forms could be transferred to α-TiOPc deposit film via evaporation deposition. Our results also indicated that the best crystallinity of α-TiOPc deposit film could be obtained from γ-TiOPc powder than the rest of crystal forms at this temperature. Moreover, the full width at half maximum (FWHM) of the peak at 7.5 degree of deposit film in the XRD spectra was 0.23, 0.27 and 0.29 for γ, α and amorphous powder, respectively. The XRD results indicated that the largest grain size of α-TiOPc deposit film was prepared from γ-TiOPc powder. The α-TiOPc deposit film prepared from amorphous TiOPc powder had the smallest grain size even the substrate temperature kept at 180°C. The XRD data of TiOPc films which prepared from α-TiOPc powder and deposited on OTS treated Si/SiO_2 surfaces at 150°C, 180°C and 220°C are shown in Figure 1b. The XRD data indicated that higher deposition temperature favors triclinic phase II (α-TiOPc). In fact, triclinic phase II TiOPc thin film can be prepared from all crystal powders at elevated temperature higher than 220°C. The FWHM of the 7.5 degree peak can achieved 0.22. The result suggested that the larger grain size of α-TiOPc deposit film can also be prepared from both α -TiOPc and amorphous TiOPc powders at higher substrate temperature.

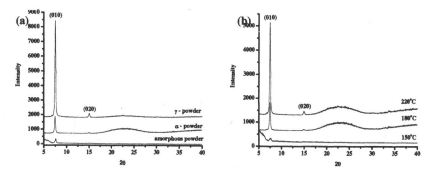

Figure 1. X-ray diffraction patterns of TiOPc films deposited on OTS treated Si/SiO$_2$ surfaces from (a) amorphous-TiOPc powder, α -TiOPc powder and γ-TiOPc powder at 180°C and (b) α - TiOPc powder at150°C, 180°C and 220°C

Morphology analysis

The morphology of TiOPc deposit films on OTS treated Si/SiO$_2$ surfaces from various TiOPc crystal forms at 180°C were observed by SEM. Highly ordered α-TiOPc films were observed in Figure 2. The α-TiOPc deposit film from γ-TiOPc powder had the biggest grain size and the α-TiOPc deposit film from amorphous TiOPc powder had the smallest grain size. The α-TiOPc film from γ-TiOPc powder was expected to have better electrical characteristics owing to the highly ordered vacuum deposited films and biggest grain size.

Figure 2. The SEM images of TiOPc films deposited on OTS treated Si/SiO$_2$ surfaces at 180°C from (a) amorphous-TiOPc powder, (b) α -TiOPc powder and (c) γ-TiOPc powder.

Electrical characteristics

The I$_{DS}$-V$_{GS}$ transfer characteristic under a drain voltage of -20 V was obtained from various TiOPc OTFTs at 180°C as shown in Figure 3. All performance factors of these TiOPc OTFTs were also summarized in Table 1, based on the measurement plots in Figure 3. The field-effect mobility (μ) in the saturated region and threshold voltage (V_{TH}) were estimated using the following equation:

$$I_D = \frac{1}{2}\frac{W}{L}C_{ox}\mu(V_G - V_{TH})^2 \tag{1}$$

where the W and L is the channel width and length, respectively, V_{TH} is the threshold voltage, μ is the field-effect mobility. C_{ox} can be only consider the capacitance of SiO_2, because of the thickness of OTS is extremely low. The highest on/off current ratio over 10^6 was obtained with a high field-effect hole mobility of 2.66×10^{-1} cm^2/Vs from a TiOPc OTFT prepared at 180 °C fabricated from γ-TiOPc powder. The excellent transfer characteristic result corresponds to the grain size effect.

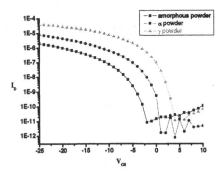

Figure 3. Drain current (I_D)-gate voltage (V_{GS}) characteristics of the TiOPc-based thin film transistors fabricated from various crystal forms: amorphous powder, α-TiOPc powder and γ-TiOPc powder.

Table 1. Performance parameters of TiOPc-based thin film transistors fabricated from various crystal forms.

Crystal form	Field-Effect Mobility (cm^2/Vs)	Threshold Voltage (V)	On/Off current ratio
amorphous-TiOPc	2.77×10^{-2}	-7.5	2.3×10^5
α-TiOPc	7.91×10^{-2}	-4.2	2.0×10^6
γ-TiOPc	2.66×10^{-1}	0.0	3.4×10^6

CONCLUSIONS

The α-TiOPc deposit film can be prepared from three different TiOPc crystal forms on the OTS to modify Si/SiO$_2$ surface. The large grain size of α-TiOPc deposit film can be prepared from γ-TiOPc crystal form power with lower substrate temperature. The highest mobility of 2.66 x 10^{-1} cm^2/V s and the highest on/off current ratio of 10^6 were obtained. The similar thin film can also be prepared by amorphous phase and α-TiOPc powders with much higher substrate at 220°C.

ACKNOWLEDGMENTS

The authors are grateful for the financial support provided by National Science Council of the Republic of China under Grant No. NSC98-2627-B-009-007-.

REFERENCES

1. C. D. Dimitrakopoulos, and P. R. L. Malenfant, "Organic thin film transistors for large area electronics," *Adv. Mater.*, **14**, 99-117 (2002).
2. Z. Bao, A. J. Lovinger, and A. Dodabalapur, "Organic field-effect transistors with high mobility based on copper Phthalocyanine," *Appl. Phys. Lett.*, **69**, 3066-3068 (1996).
3. L. Li, Q. Tang, H. Li, W. Hu, X. Yang, Z. Shuai, Y. Liu, and D. Zhu "Organic thin-film transistors of phthalocyanines," *Pure Appl. Chem.*, **80**, 2231-2240 (2008).
4. A. Tsumura, H. Koezuka, and T. Ando, "Macromolecular electronic device: field-effect transistor with a polythiophene thin fim," *Appl. Phys. Lett.*, **49**, 1210-1212 (1986).
5. C. Clarisse, M. T. Riou, M. Gauneau, and M. L. Contellec, "Field-effect transistor with diphthalocyanine thin film," *Electron. Lett.*, **24**, 674-675 (1988).
6. H. Tada, H. Touda, M. Takada, and K. Matsushige, "Quasi-intrinsic semiconducting state of titanyl-phthalocyanine films obtained under ultrahigh vacuum conditions," *Appl. Phys. Lett.*, **76**, 873-875 (2000).
7. L. Li, Q. Tang, H. Li, X. Yang, W. Hu, Y. Song, Z. Shuai, W. Xu, Y. Liu, and D. Zhu, "An ultra closely π-stacked organic semiconductor for high performance field-effect transistors," *Adv. Mater.*, **19**, 2613-2617 (2007).
8. J. Mizuguchi, G. Rihs, and H. R. Karfunkel, "Solid-state Spectra of Titanylphthalocyanine as Viewed from Molecular Distortion," *J. Phys. Chem.*, **99**, 16217-16227 (1995).

Mater. Res. Soc. Symp. Proc. Vol. 1270 © 2010 Materials Research Society   1270-II06-59

Surface Molecular Vibrations as a Tool for Analyzing Surface Impurities in Copper Phthalocyanine Organic Nanocrystals

K. Nauka, Y. Zhao, Hou T. Ng and E.G. Hanson

Hewlett-Packard Laboratories, Hewlett-Packard Company, 1501 Page Mill Road, Palo Alto, CA 94304, USA.

ABSTRACT

Surface sensitive attenuated total reflectance IR measurements have revealed small but measurable changes of certain vibrational frequencies of the surface molecules of copper pthalocyanine nanocrystals. These changes are due to electrostatic and electrodynamic interactions between the copper phthalocyanine surface molecules and external vicinal molecular species. The observed frequency shift can be changed by modifying the population of vicinal molecular entities. The observed frequency changes were confirmed by the molecular modeling of selected molecular polar species either chemically bound or vicinal to the copper phthalocyanine molecules. The observed vibrational frequency shifts can be used as an analytical tool for probing the surfaces of organic nanocrystals.

INTRODUCTION

Copper phthalocyanine (CuPc) forms exemplary organic nanocrystals with a complex crystal structure consisting of intricate arrangement of weakly interacting molecules. A CuPc molecule consists of a planar porphyrin-like structure with a central Cu atom and four peripheral benzene rings. Because of π-conjugation of the central carbon atoms, CuPc exhibits semiconductor behavior that has found a number of potential applications in organic electronics [1]. In addition, the beneficial arrangement of its molecular orbitals provides for a visible energy gap making CuPc a popular cyan pigment used in a variety of applications ranging from printing on a paper to fabric dyeing [2]. For these reasons, CuPc nanocrystals with uniform and well-controlled sizes are readily available commercially. However, various levels of impurities are typically present in these materials.

Electrostatic and electrodynamic interactions between an organic molecule and its vicinal chemical species are known to affect its basic structural and electronic properties, as in the case of solvatochromism [3]. It is conceivable that these interactions could approach the bonding strength of surface molecules within an organic crystal, like CuPc, and could modify molecular arrangement within the surface region of a crystal. This work aimed at detecting such modifications and, in particular, changes of vibrational frequencies of the surface molecules within the CuPc nanocrystals.

EXPERIMENTAL DETAILS

A number of commercially available pigment-quality CuPc materials were used in this study. They differed in terms of impurities either in the form of undesirable residues originating from the pigment manufacturing process or intentionally introduced to control properties of the pigment powder. The amount of impurities was quantified using the photoelectron spectroscopy and IR absorption [4]. In addition, pure CuPc (triply sublimated) was also studied. All samples consist of β-CuPc nanocrystallites having the shape of elongated prisms with dimensions between 50 nm and 200 nm, as confirmed by the electron microscopy.

IR absorption of the CuPc nanocrystallites was measured in transmittance mode and in attenuated total reflectance (ATR) mode. Transmittance was determined by suspending nanocrystals in an IR transparent solid medium (KBr). In the case of ATR measurements, nanocrystals were scattered on the surface of the ATR Ge crystal and gently pressed against the crystal to provide optical contact. Nanocrystals were placed on the ATR crystal as dry powder or in form of aqueous suspension with the solvent removed before the ATR measurement.

Transmittance IR measurement provides information about molecular vibrations occurring mostly within the bulk of the nanocrystals. ATR signal relies on the evanescent wave penetrating a relatively shallow region below the surface of a nanocrystal. In addition, only the nanocrystals with the main surfaces parallel or close-to-parallel to the ATR crystal surface could contribute to the ATR signal. The surface region of a nanocrystal can be defined as region having thickness corresponding to distance between the Cu atom in the center of CuPc molecule and its outermost H atom. Assuming CuPc absorption coefficient to be equal approximately 10^5 cm^{-1} and using the known formula [5] describing the decay of an evanescent wave as it penetrates the nanocrystal, it can be estimated that between 2% and 5% of the collected ATR signal originated from the surface region of CuPc nanocrystal. The remaining part of the ATR signal is due to bulk molecular vibrations. The resulting signal, consisting of these two components, may have its peak energy slightly shifted when compared to the corresponding transmittance absorption frequency.

The main advantage of using nanocrystals in this experiment stems from their high surface-to-bulk ratio and the corresponding relatively large contribution of the crystal's surface region to the overall ATR signal. However, even when organic nanocrystals are used the observed changes of the vibrational frequencies are small and can be difficult to measure. Therefore, they require highly sensitive ATR-FTIR instrument capable of resolving very small changes of the vibrational frequencies. The IR absorption was measured with a research grade FTIR apparatus using the highest available spectral resolution (< 0.1 cm-1), averaging 200 scans, repeating the same measurement multiple times in order to minimize measurement errors, and finally fitting the top part of the experimental curve with a Gaussian function to determine the exact position of absorption maximum.

Commercial grade CuPc frequently contains intentional molecular impurities that are in a vicinal position with respect to the crystal's surface and are either chemisorbed or physisorbed to the crystal's surface molecules [2]. Their rearrangement (or removal) should modify forces that they impart upon the surface molecules. Therefore, selected CuPc nanocrystals were modified by Soxhlet extraction using solvents with different polarities (hexane, acetone). The extraction caused measurable removal of at least some of the impurities, as confirmed by weight loss measurement (typical weight loss of the order of 0.1% to 1%). Surface sensitive, glancing-angle

X-ray diffraction confirmed that extraction did not affect crystalline structure of the CuPc nanocrystals.

DISCUSSION

The origin of IR absorption features of the CuPc is now well understood [6] and some of the major absorption features are identified in Figure 1. The strongest absorption peaks that do not overlap with other absorption maxima are particularly well suited for the aforementioned analysis. Figure 2 presents ATR and transmittance spectra of a particular vibration (out-of-plane bending of the C-H bond within the peripheral benzene ring) occurring at 733.5 cm^{-1}. Comparison of the as-received CuPc with the same material after hexane and acetone extraction showed small but measurable changes of the vibration frequency as measured by ATR, while in the corresponding transmittance measurement the absorption frequency remained the same. Results of this experiment are summarized in Table 1.

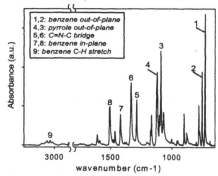

Figure 1. FTIR transmission spectrum of CuPc.

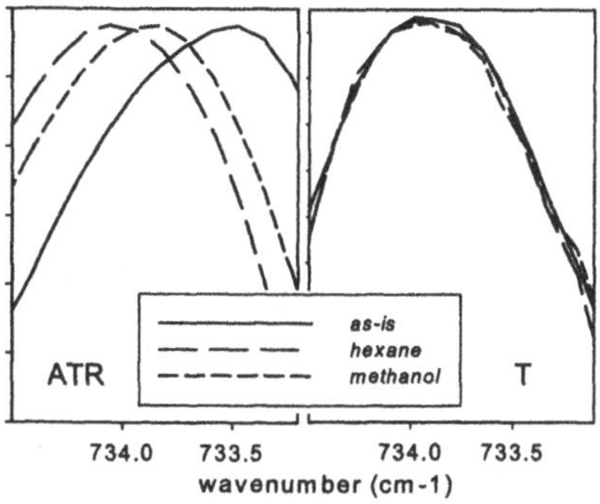

Figure 2. Benzene out-of-plane (peak 1) vibrational absorption measured by ATR and by transmittance (T).

The observed upward shift of the ATR measured vibrational frequency is likely due to decrease in the interactions between the peripheral benzene rings and the vicinal molecular species. Interestingly, vibrational frequency after extraction, that removed at least some of the impurities, was closer to the corresponding bulk vibrational frequency as measured by transmittance analysis. It was also close to ATR measured vibrational frequency of a high purity CuPc (Table 1).

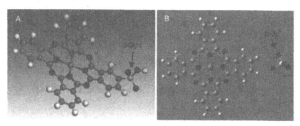

Figure 3. Molecular modeling: A. DFT calculations (PBE / DMOL3 and VAMP - Materials Studio) (Accelrys Inc.); B. Time-dependent DFT calculations (MO6L - DZQ).

The effect of molecular interactions between the CuPc and a vicinal molecular impurity was further elucidated with the help of molecular modeling. Figure 3 presents the case when a sulfonate group is chemically bonded to a CuPc molecule or when a sulfate anion is present at its minimum energy position in the vicinity of a CuPc molecule. This calculation does not represent interactions between the surface of a CuPc nanocrystal and a vicinal molecular specie (only one CuPc molecule is used) but provides a reference for the postulated explanation of the experimental results. A tentative agreement between the experimental data and the corresponding calculations supports our explanation. Their discrepancy is likely due to inability to model interactions between the vicinal molecules and an extended CuPc molecular crystal with size corresponding to nanocrystals used in the experiment.

Table 1. Measured and calculated vibrational frequencies (benzene out-of-plane / 733.5 cm^{-1})

CuPc sample	measured				calculated		
	ATR as-is	ATR hexane	ATR methanol	T as-is	CuPc molecule	SO$_3$H bonded	SO$_4^{2-}$ vicinal
wavenumber (cm^{-1})	733.50	734.05	734.0	733.95	732.4	729.1	719.4

Comparison of other molecular absorption maxima and, in particular, strong pyrrole in-plane modes occurring at around 1100 cm^{-1}, demonstrated a similar shift of the ATR measured vibration frequencies as in the case of 733. 5 cm^{-1} benzene out-of-plane vibration. However, the error of these measurements is larger than in the case of 733. 5 cm^{-1} mode due to partial peak overlap and weaker overall absorption. Further detailed analysis is needed to ascertain details of the frequency shifts for these vibrational modes.

CONCLUSIONS

It was observed that interactions between the surface molecules of the CuPc nanocrystals and vicinal molecular impurities can lead to frequency shift in the absorption spectrum of the surface molecules. This shift, though small, can be measured with help of high spectral resolution ATR FTIR. Magnitude of the frequency shift appears to correlate with the type of vicinal species. Thus, ATR FTIR measurement could be potentially used as an analytical tool to provide information about the surface properties of organic crystals.

REFERENCES

1. Z. Bao, A.J. Lovinger, A. Dodabaladur, *Appl.Phys.Lett.* **69**, 3066 (1996).
2. H. Zollinger, *Color Chemistry*, Wiley-VCH & VHCA, Zürich 2003.
3. C.Reichardt, *Chem.Rev.* **94**, 2319 (1994).
4. K. Nauka, H.T. Ng, E.G. Hanson, to be published in the MRS Proc.Ser.Vol.1270 (2010).
5. W. Suětaka, *Surface Infrared and Raman Spectroscopy*, Plenum Press, New York 1995.
6. R.A. Jones, G.P. Bean, *The Chemistry of Pyrroles*, Academic Press, London 1977.

Mater. Res. Soc. Symp. Proc. Vol. 1270 © 2010 Materials Research Society — 1270-II06-68

Effect of film morphology on charge transport in C_{60} based Organic Field Effect Transistors

Mujeeb Ullah[1], A. K. Kadashchuk[2,3], P. Stadler[4], A. Kharchenko[5], A. Pivrikas[4], C. Simbrunner[1], N. S. Sariciftci[4] and H. Sitter[1]

[1]Institute of Semiconductor and Solid State Physics,
Johannes Kepler University Linz, Austria.
[2]IMEC, Kapeldreef 75, B-3001, Heverlee, Belgium.
[3]Institute of Physics, National Academy of Sciences of Ukraine,
Prospect Nauky 46, 03028 Kyiv, Ukraine
[4]Linz Institute of Organic Solar Cells (LIOS),
Johannes Kepler University Linz, Austria.
[5]PANalytical B.V, Lelyweg 1, P.O.Box 13, 7600 AA Almelo, The Netherlands.

ABSTRACT

The critical factor that limits the efficiencies of organic electronic devices is the low charge carrier mobility which is attributed to disorder in organic films. In this work we study the effects of active film morphology on the charge transport in Organic Field Effect Transistors (OFETs). We fabricated the OFETs using different substrate temperature to grow different morphologies of C_{60} films by Hot Wall Epitaxy. Atomic Force Microscopy images and XRD results showed increasing grain size with increasing substrate temperature. An increase in field effect mobility was observed for different OFETs with increasing grain size in C_{60} films. The temperature dependence of charge carrier mobility in these devices followed the empirical relation named as Meyer-Neldel Rule and showed different activation energies for films with different degree of disorder. A shift in characteristic Meyer-Neldel energy was observed with changing C_{60} morphology which can be considered as an energetic disorder parameter.

INTRODUCTION

The field of organic field-effect transistors (OFETs) has gained considerable attention because of its potential for emerging flexible electronics and active matrix backplanes for displays [1]. Considerable efforts have been devoted to increase the performance of OFETs [2, 3] which critically depends on charge carrier transport properties in organic semiconductor layers. Although the charge mobility in OFETs has been improved significantly in the last years, it still remains by far lower than that of conventional inorganic devices and, along with stability issues, makes a current bottleneck for the large scale industrial application of organic electronic devices. Charge mobility is primarily determined by the disordered nature of conventional thin organic films, therefore methods capable of improving organic thin film morphologies are thus of considerable scientific as well as technological interest.

In amorphous disordered thin organic films the electronic states are localized and energetically distributed, so that transport occurs via incoherent thermally-assisted hopping [4-6] in the density-of-states (DOS) distribution most often described by a Gaussian disorder model of Bässler [4].

$$\mu_{(T,E)} = \mu_0 \exp\left[-\left(\frac{2}{3}\hat{\sigma}\right)^2\right] \exp\left[-C\left(\hat{\sigma}^2 - \Sigma^2\right)E^{1/2}\right] \qquad (1)$$

$\hat{\sigma} = \sigma / k_B T$ and Σ are parameters characterizing energetic disorder and positional disorder, where σ is the width of the Gaussian density of states, E is the electric field, μ_0 is a mobility prefactor related to the zero-field mobility at infinite temperature and in the absence of disorder, and C is an empirical constant.

Recently it was recognized that charge carrier mobility in actual organic electronic devices depends essentially on carrier concentration which was theoretically rationalized within Extended Gaussian Disorder model [7-9]. In the latter case a sizeable fraction of the DOS distribution is occupied by charge carriers, so carrier jumps from the Fermi level dominate the charge transport giving rise to an Arrhenius-type $\ln(\mu) \propto T^{-1}$ temperature dependence of the mobility [10] with virtually constant (yet dependent on carrier concentration) activation energy reflecting the temperature independent position of the Fermi level position with respect to the center of the DOS. This is in contrast to the case of low carrier concentration, characterized by the non-Arrhenius type temperature dependence $\ln(\mu) \propto T^{-2}$ in the Bässler formula since in such a case the charge transport is dominated by hopping from the equilibrium occupational DOS distribution which is also temperature dependent [4, 6].

In a number of studies [11-13] is has been well documented that charge carrier mobility obeys an Arrhenius-type $\mu(T)$ dependence. Measured at different gate voltages (V_G) and, concomitantly, at different charge carrier densities, these Arrhenius plots intersect at a given finite temperature T_0, suggesting that the Meyer-Neldel rule (MNR) [14] is obeyed. The MNR is an empirical relation, originally derived from chemical kinetics [14]. More generally, it states that the MNR can occur in any situation which involves an thermally activated process when an increase of the activation energy E_a is partially compensated by an increase of the prefactor. Thus the thermally activated charge carrier mobility in disordered organic semiconductors is empirically given by

$$\mu = \mu_{p0} \exp\left[-E_a\left(\frac{1}{k_B T} - \frac{1}{E_{MN}}\right)\right], \qquad (2)$$

where $E_{MN} = k_B T_0$ is called the 'Meyer-Neldel energy' and μ_{p0} is a constant prefactor. E_a is the activation energy being affected by changing the charge carrier concentration, for instance, by changing the gate voltage in OFETs. Understanding of the MNR effect for the charge carrier mobility and the physical meaning of E_{MN} is still heavily under debate and several theoretical attempts based on polaron concept have been suggested to justify the MNR in different systems [15].

Most recently, Fishchuk et al. [9, 16] have formulated an analytical theory based on the Effective Medium approximation (EMA) to describe the Meyer-Neldel compensation rule for the FET mobility of charge carriers irrespective of their polaronic character by employing the conventional hopping transport concept for a disordered system with a Gaussian DOS distribution and Miller-Abrahams jump rates. The MNR effect was recovered regarding the temperature dependences of the charge carrier mobility upon varying the carrier concentration (gate voltage), but not regarding the variation of the width of the DOS, σ [16]. It was also shown that the results of the EMA calculations can be parameterized in terms of an approximate analytical equation for the charge carrier mobility μ_e as a function of $\sigma / k_B T$, the ratio of densities of occupied and total transport states n/N, and the ratio a/b of the intersite distance (a) and the localization radius (b) of the charged site. At moderately high temperatures and commonly accepted value $a/b = 10$ [7, 8] for organic semiconductors, the approximate relation for the temperature dependent FET mobility reduces to

$$\frac{\mu_e}{\mu_0} \propto \exp\left[-E_a \left(\frac{1}{k_B T} - \frac{1}{k_B T_0} \right) \right], \tag{3}$$

where

$$E_a = \left[0.75 - 0.67 \log_{10}\left(\frac{n}{N} \right) \right] \times \sigma \quad \text{and finally} \qquad T_0 = \frac{2}{5} \frac{\sigma}{k_B}. \tag{4}$$

Eq.(3) is nothing else than the conventional Meyer-Neldel relation (cf. Eq. (2)). E_a depends on the relative carrier concentration n/N and on the energetic disorder parameter σ, so it can be influenced by both parameters. An important prediction of the suggested theory [16] is that the MNR energy E_{MN} in organic semiconductors is directly related to the width of the Gaussian DOS, σ, providing thus a new method for evaluation of the amount of the energetic disorder in the material.

RESULTS

In the present work, we performed a systematic experimental study of the interrelation between the energetic disorder and the MNR energy as derived from the temperature dependent FET mobilities in C_{60} films grown at different conditions resulting in different film morphologies. The obtained results present a direct experimental verification of the suggested theoretical model by Fishchuk et al. [16] for the Meyer-Neldel compensation rule in organic FETs and demonstrate the impact of the resulting film quality on the MNR behavior of the FET mobility.

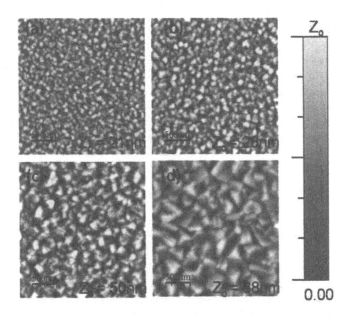

Fig.1. AFM images of C_{60} films grown at different substrate temperatures (a) at $130°$ C without preheating (b) at $150°$ C with preheating at $155°$ C for 20 mins. (c) at $200°$ C with preheating at $205°$ C for 20 mins. (d) at $250°$ C with preheating at $255°$ C for 20 mins.

The top contact-bottom gate OFET devices were fabricated on ITO covered glass substrates using divinyltetramethyldisiloxane-bis(benzocyclobutane) (BCB) as gate-insulating layer. The thin C_{60} films were deposited at different substrate temperatures using standard flush evaporation at room temperature or Hot-Wall Epitaxy (HWE) technique. These films were polycrystalline in nature as shown from the AFM images in Fig.1 and XRD results from Fig.2. As one can see from Fig.1 the film morphology depends quite considerably on the growth conditions – increasing the substrate temperature (T_{sub}) results in larger grain size of the C_{60} films. The smallest grains were obtained in films grown at room temperature by flash evaporation without pre-heating the substrate and the largest ones in films grown by HWE at $T_{sub} = 250$ C°. The XRD results also confirm the increasing crystalinity in the C_{60} films with increasing substrate temperature as shown in Fig.2. Finally LiF/Al top contacts were evaporated in high vacuum on top of the C_{60} films to complete the devices.

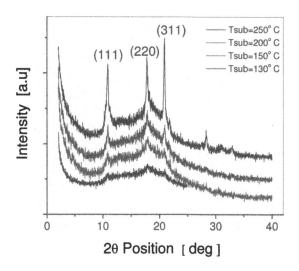

Fig.2. XRD results for C_{60} films grown by HWE on BCB at different substrate temperatures.

The completed devices were loaded for electrical characterization under nitrogen atmosphere in an Oxford cryostat and all measurements were carried out at a vacuum of around 10^{-6} mbar and were recorded by Agilent 2000 SMU. The temperature was changed in the range from 320K to 80K with a step of 20 K. The field-effect mobility μ_{FE} of the C_{60}-based OFET has been determined in the linear regime of the $I_D - V_g$ characteristics (at low source-drain voltage $V_D = 2$ V). The applied source-gate electric field in this regime was much larger than the in-plane source-drain field, which resulted in an approximately uniform density of charge carriers in the conductive channel.

Fig.3 (symbols) shows the field-effect mobility μ_{FE} as a function of inverse temperature measured for moderately large temperatures at different gate voltages V_G in the C_{60} films grown by HWE at $T_{sub} = 130°$ C and $250°$ C (Fig. 1a and d, respectively). It is evident that the extrapolation of these graphs intersect at the isokinetic temperature $T_0 = 351$ K and 250 K for the films grown at $130°$ C and $250°$ C, respectively, thus clearly demonstrating a MNR-type behavior. This yields in corresponding MNR energies of $E_{MN} = 30.2$ and 21.5 meV (Fig. 3). The MNR effect has been reported recently by our team for the flush evaporated C_{60} film which was deposited at room temperature [16], which featured also a clear MNR-type behavior resulting in $E_{MN} = 34$ meV for this highly disordered film.

(a)   (b)

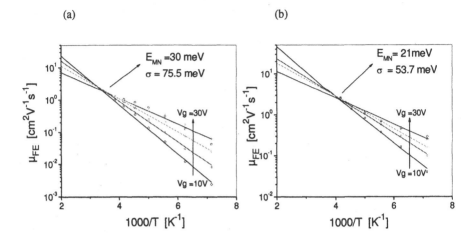

Fig.3. Temperature dependence of the mobility in C_{60} OFETs (a) with C_{60} film grown at substrate temperature of $130°$ C and (b) C_{60} film grown at substrate temperature of $250°$ C.

DISCUSSION

The results presented in Fig. 3 are in excellent agreement with the predictions of the recent EMA theoretical model [16] as described above. The solid curves in Fig. 3a and 3b represent a linear fit of the experimental data using Eq.(2) which provides T_0. Using Eq. (3) for each T_0, determined from experiment, the corresponding width of the DOS σ =75.5 meV and 53.7 meV can be obtained for films grown at $130°$ C and $250°$ C, respectively. For C_{60} films thermally evaporated $\sigma \cong 88 meV$ was determined [16]. This suggests that the energetic disorder in C_{60} films decreases substantially with increasing substrate temperature being accompanied with an enhancement of the mobility, which also correlates with an increasing grain size in the films. Fig. 4 summarizes the mobility data at room temperature, together with the E_{MN} as the function of the grain size of the C_{60} layers, deduced from AFM picture. It can be clearly seen, that with increasing grain size the degree of disorder decreases and the mobility increases. The weakening of the $\mu_e(T)$ dependence for the C_{60} film grown at higher substrate temperature (Fig. 3) is a direct evidence for the reduced energetic disorder in this film.

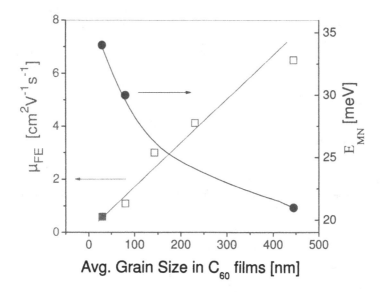

Fig.4. Field effect mobility and Meyer-Neldel energy as function of grain size in C_{60} films.

It is worth noting that $E_{MN} = 21.5$ meV observed for the C_{60} film grown by HWE at 250 K is probably the smallest reported so far for FET mobility in organic transistors and suggest a very reduced energetic disorder with $\sigma = 53.7$ meV. However, it is similar to the value of 50 meV determined by time-of-flight (ToF) measurements in MeLPPP conjugated polymer [17]. Normally, σ-values determined for the FET mobility are somewhat larger than that for the ToF mobilities measured in the same material due to the interface effects in OFETs (surface traps, surface dipoles, etc.) [18, 19]. The observed considerable change in E_{MN} is amazing as previous studies pointed out [11] that this energy for different materials appears to be always surprisingly close to 40 meV. In the present study an organic BCB non-polar gate isolator was used, so the interface effects are expected to be very weak compared to say SiO_2. This circumstance combined with the optimized structure of C_{60} films grown by HWE technique can explain the reduced energetic disorder in the C_{60} films.

In conclusion – Meyer-Neldel behavior of the temperature dependent FET mobility has been studied in C_{60} films grown at different growth conditions and the experimental observations are found to be in a nice agreement with the predictions of the recently suggested theoretical model for organic semiconductors with a Gaussian DOS distribution. The apparent MNR effect was observed at variable activation energy of the mobility due to changing the gate voltage and not due to changing the energetic disorder, exactly as predicted by the theory above. An amazing strong shift of the Meyer-Neldel (isokinetic) temperature T_0 (from $T_0 = 408$ K $\to$ 351 K $\to$ 250

K), was found in C_{60} films upon increasing the substrate temperature during film growth from room temperature to T_{sub} =130° C and T_{sub} =250° C, respectively, and this correlates with the change of the film morphology and, consequently, a change of the energetic disorder. The latter depends substantially on the C_{60} film morphology having an impact on charge mobility value; the latter correlates with the average grain size of the grown C_{60} films. An unusually small MNR energy of 21.5 meV found for the C_{60} films grown at T_{sub} =250° C correlates with a reduced energetic disorder with σ =53.7 meV and, consequently, very high charge mobility in these films.

Acknowledgement

The research was implemented within the bilateral ÖAD Project UA-10/2009 and supported by the Ministry of Education and Science of Ukraine (project M/125-2009) and the Austrian Science Foundation (NFN projects S9706, S9711) and by the STCU project no. 5258.

References:

1. H. F. A. Huitema, et al.. *Nature* 414, 599 (2001), and G. H. Gelinck, et al., *Nat. Mater.* 3, 106 (2004).
2. H. Sirringhaus *et al.*, Nature (London) **401**, 685 (1999).
3. S. R. Forrest, Nature (London) **428**, 911 (2004).
4. H. Bässler, Physica Status Solidi B 175, 15 (1993).
5. P.W.M. Blom, and M.C.J.M. Vissenberg, *Mater. Sci. Eng.* 27, 53 (2000).
6. V. I. Arkhipov, et al, in *Semiconducting Polymers: Chemistry, Physics and Engineering, 2nd Edition"*, Eds. G. Hadziioannou and G. Malliaras, Wiley-VCH Verlag, Weinheim, (2007).
7. W.F. Pasveer, et al., Phys. Rev. Lett. 94, 206601 (2005).
8. R. Coehoorn, W. F. Pasveer, P. A. Bobbert, and M. A. J. Michels, Phys. Rev. B **72**, 155206 (2005).
9. I. I. Fishchuk, et al., Phys. Rev. B, 76, 045210 (2007).
10. N. I. Craciun, J. Wildeman, and P.W. M. Blom, Phys. Rev. Lett. **100**, 056601 (2008).
11. E. J. Meijer, E. J. Meijer, M. Matters, P. T. Herwig, D. M. de Leeuw, and T. M. Klapwijk, Appl. Phys. Lett., **76**, 3433 (2000).
12. J. Paloheimo and H. Isotalo, Synth. Met. **55**, 3185 (1993).
13. Mujeeb Ullah, T. B. Singh, H. Sitter, N. S. Sariciftci, Appl. Phys. A, 97, 521 (2009).
14. W. Meyer and H. Neldel, Z. Tech. Phys. **18**, 588 (1937).
15. A. Yelon and B. Movaghar, Phys. Rev. Lett. 65, 618 (1990); D. Emin, Phys. Rev. Lett., 100, 166602 (2008).
16. I. I. Fishchuk, A. Kadachchuk, J. Ganoe, Mujeeb Ullah, H. Sitter, Th. B. Singh, N. S. Sariciftci, and H. Bässler, Phys. Rev. B, **81**, 045202 (2010).
17. H. Sitter, T. Nguyen Manh, D. Stifter, *J. Cryst. Growth 174*, 828 (1997).
18. D. Hertel, H. Bässler, U. Scherf, and H. H. Hörhold, *J. Chem. Phys.* 110, 9214 (1999).
19. S. C. Veenstra and H. T. Jonkman, J. Polym. Sci., Part B: Polym.Phys. **41**, 2549 (2003).
20. B. N. Limketkai and M. A. Baldo, Phys. Rev. B **71**, 085207 (2005).

Mater. Res. Soc. Symp. Proc. Vol. 1270 © 2010 Materials Research Society 1270-II06-93

Poly(3-hexylthiophene) Nanofibers Fabricated by Electrospinning and Their Optical Properties

Surawut Chuangchote, Michiyasu Fujita, Takashi Sagawa[*], and Susumu Yoshikawa[*]
Institute of Advanced Energy, Kyoto University, Gokasho, Uji, Kyoto 611-0011, Japan
E-mail: t-sagawa@iae.kyoto-u.ac.jp; s-yoshi@iae.kyoto-u.ac.jp

ABSTRACT

Beaded fibers and/or uniform, smooth-surface fibers of conductive polymers with the average diameters ranging in nanometers to sub-micrometers were fabricated by electrospinning of a mixture of poly(3-hexylthiophene) (P3HT) and polyvinylpyrrolidone (PVP) in a mixed solvent of chlorobenzene and methanol. After the removal of PVP from as-spun fibers by Soxhlet extraction, pure P3HT fibers were obtained as a spindle-like with groove-like morphological appearance which may be widely applicable for some specific applications, such as photovoltaic cells, thin film transistors, and light emitting diodes. Optical properties, including UV absorption and photoluminescence (PL) of fibers were investigated. As-spun fibers showed relatively higher conjugation length and different chain distribution, in comparison with the cast film.

INTRODUCTION

Since the discovery of the electrical conductivity in π-conjugated polymers thirty years ago [1], conductive polymers have become the focus of intense development and research activities around the world. Their use as organic semiconductors and synthetic metals in various optical and electrical applications has led to rapid growth of the field. Many different families of conjugated polymers are being studied. Among them, polythiophene and its derivatives represent an interesting family because of their nonlinear optical characteristics, photo- and electroluminescent properties, and charge-carrier mobility, which make them attractive for use in applications that include organic light-emitting diodes, transistors, and especially solar cells [2-3].

One-dimensional characteristic-ultrafine fibers have been interested recently because when the diameters of polymeric materials are shrunk to sub-micrometers or nanometers, there appear to be several specific characteristics such as improved mechanical performance, very large surface area to volume ratio, and flexibility in surface functionalities [4]. Among various fibers processing technique, electrospinning has become one of simple techniques, which uses electrostatic forces to produce polymeric, ceramics, and composite continuous ultrafine fibers with diameters ranging from microns down to a few nanometers [4].

Recently, we have reported the fabrication of conductive polymer nanofibers of poly[2-methoxy-5-(2'-ethylhexyloxy)-1,4-phenylene-vinylene] (MEHPPV), by electrospinning with blending of poly(vinyl pyrrolidone) (PVP) [5-7]. Ultrafine MEH-PPV fibers could be obtained from electrospinning and subsequent Soxhlet extraction. Obtained fibers were applied to organic photovoltaic cells [8]. We here report an attempt to apply this technique to fabricate ultrafine nanofibers of poly(3-hexylthiophene) (P3HT). Optical properties, including UV-vis absorption and photoluminescence (PL) of the fibers are also reported.

EXPERIMENT

Solutions of P3HT were prepared in the mixed solvent of chlorobenzene and methanol (85:15 v/v) under vigorously stirring. The obtained solutions were sonicated for 10 min and heat to $40\,^{\circ}C$ to exceed the dissolubility of P3HT. A controlled amount of PVP was added into P3HT solutions with vigorously stirring.

As-prepared solutions were electrospun by loading each of them in a 3-ml plastic syringe with a 60°C-heated jacket. The nozzle was a blunt-end stainless-steel gauge 22 needle. The collector was a sheet of aluminum foil on a plastic plate. A high voltage power supply was used to charge the solution across an electrode in solution and the collector. As-prepared solutions were electrospun under an applied electrical potential of 15 kV over a fixed collection distance of 15 cm at room temperature. The solution feed rate was controlled at $1\ mL\cdot h^{-1}$. The collection time were fixed for all experiments at 1 min. For the removal of PVP from as-spun P3HT/PVP fibers, the Soxhlet extraction was carried out. Methanol was used as a solvent at the extraction temperature of 75°C. The extraction time was fixed at 12 hr. The samples were then dried overnight at 60°C in a vacuum oven. The morphological appearance of the as-spun fiber mats was investigated by a scanning electron microscope (SEM), operating at an acceleration voltage of 10 kV. UV-Vis and PL spectra were investigated.

RESULTS AND DISCUSSION

Fabrication of P3HT Nanofibers

In the simple electrospinning technique of two polymers blended without any specially designed spinneret, clear dissolved solution from the mixed polymers should be prepared. In this work, P3HT was expected to blend with PVP, an easily spinnable and easily extractable polymer, in mixed solvent. P3HT was found to be well dissolved in chlorobenzene, while electrospinning of PVP solutions in methanol with various concentrations resulted in formation of ultrafine fibers. Therefore, the mixed solvent of chlorobenzene and methanol was chosen for electrospinning of P3HT/PVP. Electrospinning of blended solutions resulted in ultrafine P3HT/PVP composite fibers with the average diameters ranging in sub-micro- down to nanometers (2 μm down to 60 nm) (see an example of the fibers obtained in Figure 1(a)). The average diameter of the as-spun fibers decreased and the color of as-spun fibers changed in darkness of violet-red with decreasing the concentration of P3HT or PVP.

After the removal of PVP from as-spun fibers by Soxhlet extraction, pure P3HT fibers were obtained as a ribbon-like structure aligned with wrinkled surface in fiber direction (see Figure 1(b)). These results suggest that there was a phase separation of P3HT and PVP in the sub-micrometer scale occurred during instability and the confinement and electric field during electrospinning could enhance the orientation of this phase separation or also might be that of polymer chains in fiber direction [5-6]. In addition, the removal of PVP also resulted the decreasing the diameter of fibers and thickness of fiber mats, and the enhancement of contacts parts of fiber mats.

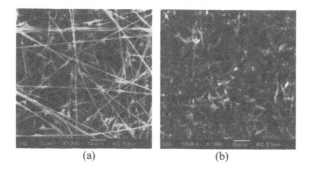

| (a) | (b) |

Figure 1. SEM images (scale bar = 10 μm) of (a) as-spun P3HT/PVP composite fibers and (b) P3HT fibers (after extraction of PVP from the composite fibers). Composite fibers were electrospun from solutions of 6.0% w/v P3HT/PVP (P3HT:PVP = 2:4 w/w) in chlorobenzene/methanol (85:15 v/v). The applied electrical potential was 15 kV over the fixed collection distance of 15 cm.

Optical Properties of P3HT Nanofibers

Figure 2 shows the normalized UV-vis absorption spectra from cast film and electrospun fibers of both P3HT/PVP and P3HT (after removal of PVP from the composite in case of fibers).

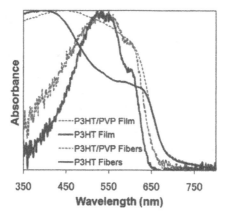

Figure 2. UV-vis spectra of P3HT/PVP and P3HT in two forms of cast film and electrospun fiber (P3HT:PVP = 1:3 for blended samples).

The absorption peak and shoulder of P3HT and P3HT/PVP film appeared at quite same wavelength. However, the absorbance band of P3HT/PVP cast film is broader than P3HT one, suggesting a more inhomogeneous environment of P3HT in P3HT/PVP cast film. The absorption peak and shoulder of P3HT/PVP fibers showed the remarkable shift compared to the cast film implies that there is a change in chain distribution within each fiber towards a better delocalized π-conjugation and more extended conformation [9]. The stretching of the liquid jet during bending instability of electrospinning should be responsible for the increasing in conjugation length. After the removal of PVP from P3HT/PVP fibers, P3HT fibers showed the relatively shift as compared to P3HT/PVP fibers indicates the change in chain distribution after disappear of PVP.

For the PL spectra (Figure 3), with an excitation wavelength of 520 nm, P3HT/PVP and P3HT films showed one emission peak at the wavelength of about 640 nm. A relatively red-shift was observed from the emission of P3HT film compared with P3HT/PVP film. It indicates that there might be the increase in aggregation of P3HT phase in micro scale. For as-spun P3HT/PVP and P3HT fibers, compared with the cast films, it showed significant red-shift in emission peaks. Same as the case of the films, a red-shift was observed from the emission of P3HT fibers compared with P3HT/PVP fibers (650 nm and 655 nm, respectively). This result corresponds to UV-vis spectra of film and fibers, showing remarkable difference in terms of delocalized π-conjugation and chain distribution.

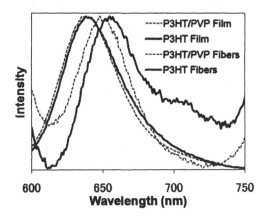

Figure 3. PL spectra of P3HT/PVP and P3HT in two forms of cast film and electrospun fiber (P3HT:PVP = 1:3 for blended samples).

CONCLUSIONS

Conductive polymers with the average diameters ranging in nanometers to sub-micrometers were fabricated by electrospinning of a mixture of poly(3-hexylthiophene) (P3HT) and polyvinylpyrrolidone (PVP). After the removal of PVP from as-spun fibers by Soxhlet extraction, pure P3HT fibers were obtained as a spindle-like with groove-like morphological appearance which may be widely applicable for some specific applications, such as photovoltaic cells, thin film transistors, and light emitting diodes. Optical properties, including UV absorption and photoluminescence (PL) of fibers were investigated. As-spun fibers showed relatively higher conjugation length and different chain distribution, in comparison with the cast film.

ACKNOWLEDGMENTS

This work was supported by grant-in-aids from Japan Society for the Promotion of Science (JSPS) under the JSPS Postdoctoral Fellowship for Foreign Researchers.

REFERENCES

1. H. Shirakawa, E. J. Louis, A. G. MacDiarmid, C. K. Chiang, and A. J. Heeger, *J. Chem. Soc. Chem. Comm.* **16**, 579 (1977).
2. H. Hoppea and N. S. Sariciftci, *J. Mater. Res.* **19**, 1924 (2004).
3. F. Padinger, R. S. Rittberger, and N. S. Sariciftci, *Adv. Funct. Mater.* **13**, 85 (2003).
4. S. Chuangchote, A. Sirivat, and P. Supaphol, *Nanotechnology* **18**, 145705 (2007).
5. S. Chuangchote, T. Sagawa, and S. Yoshikawa, *Jpn. J. Appl. Phys.* **47**, 787 (2008).
6. S. Chuangchote, T. Sagawa, and S. Yoshikawa, *Macromol. Symp.* **264**, 80 (2008).
7. S. Chuangchote, T. Sagawa, and S. Yoshikawa, *Mater. Res. Soc. Symp. Proc.* **1091E**, 1091-AA07-85 (2008).
8. S. Chuangchote, T. Sagawa, and S. Yoshikawa, *Mater. Res. Soc. Symp. Proc.* **1149E**, 1149-QQ11-04 (2008).
9. B. J. Schwartz, *Annu. Rev. Phys. Chem.* **54**, 141 (2003).

Devices II

Mater. Res. Soc. Symp. Proc. Vol. 1270 © 2010 Materials Research Society   1270-II07-03

Inkjet Printed P3HT:PCBM Solar Cells: A New Solvent System Approach

Alexander Lange, Michael Wegener, Christine Boeffel, Bert Fischer and Armin Wedel
Fraunhofer Institute for Applied Polymer Research, Geiselbergstrasse 69
Potsdam-Golm, 14476, Germany

ABSTRACT

Inkjet printing and spin coating were used to deposit polymer passive layers and polymer:small molecular active layers consisting of poly(3-hexyl thiophene) and phenyl-C61 butyric acid methyl ester for organic solar cells. An inkjet solvent system consisting of chlorobenzene and trichlorobenzene was used where the ink's surface tension and viscosity were 32.5 mN/m and 1.0 to 1.5 mPa*s, respectively. Devices with inkjet printed passive and active layers were found to have slightly higher efficiency values with respect to solar cells with spin coated layers after pre-annealing. Additionally, it was determined that the achievable open circuit voltage of solar cells with inkjet printed or spin coated passive and active layers was within the range of 0.20 to 0.55 V.

INTRODUCTION

Organic solar cells based on poly(3-hexyl thiophene) (P3HT) and phenyl-C_{61} butyric acid methyl ester (PCBM) offer advantages over inorganic devices because they can be processed from solution with techniques such as spin coating and printing [1-4], the substrates used for these devices can be rigid or flexible [5] and they offer partial transparency [6]. When P3HT is mixed with PCBM, a bulk heterojunction between the two materials is formed which acts to split electron-hole pairs, or excitons. In order to generate a current in an external circuit, free holes and electrons from split excitons must be transported to the anode and cathode of the solar cell, respectively. Charge transport is typically enhanced when an optimum degree of phase separation occurs between the donor and acceptor. In general, a solar cell consists of four layers, the anode, hole transport layer (HTL), active layer and cathode, on top of a rigid or flexible substrate. Each layer has a specific function in the solar cell where the active layer absorbs incident light and generates free charges, the HTL provides a smooth and homogeneous interface between the active layer and the anode and the electrodes collect free holes and electrons. Figure 1 shows the layers of a P3HT:PCBM solar cell which are needed to successfully convert light into electricity.

Figure 1. Traditional solar cell structure which shows the two polymer layers (HTL and active layer) in addition to the electrodes and a glass substrate. Please note that HTL stands for the hole transport layer.

Due to the low thicknesses of the HTL and active layer, which together can be less than 300 nm, deposition technologies with a high degree of control and precision are needed. Most solar cell research focuses on depositing the HTL, active layer and cathode of organic solar cells because substrates with patterned anodes are commercially available. Traditionally, spin coating has been used to deposit the HTL and active layers where a centrifugal force rotates the substrate and produces thin, homogeneous films. Solar cells based P3HT:PCBM which were made with this technology have demonstrated power conversion efficiencies (η_{PCE}) of up to 5% [1]. However, the majority of the polymer solution which is used to form films during spin coating is wasted because of the centrifugal forces which rotate the substrate. For large scale production, different technologies are needed where material waste is limited and a high degree of control over the process is possible.

Inkjet printing offers a favorable alternative to spin coating because all of the material that leaves the print head is used to produce the respective layer. In addition, user-defined images which result in patterned films can be printed whereas films prepared by spin coating have no structure after preparation. Reports which demonstrated inkjet printed P3HT:PCBM solar cells have shown that printed devices can reach η_{PCE} of around 3% [2-4]. Figure 2 shows flow diagrams for spin coating and inkjet printing which detail the different steps of each process with respect to preparation of the P3HT:PCBM layer.

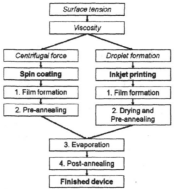

Figure 2. Flow diagrams of the active layer preparation process which utilizes spin coating or inkjet printing to generate the thin film. Useful information needed for both processes and the film formation forces used during each method are also shown (*italic*).

During spin coating, a centrifugal force is used to spread a fluid across a substrate. The rotational force causes solvent to evaporate from the fluid which results in a film with a distinct polymer:small molecule morphology. In order to obtain more defined regions of P3HT and PCBM, which can lead to percolation pathways to the anode and cathode, the active layer can be thermally treated before evaporation of the metal cathode [7]. This process is typically called pre-annealing. To further optimize the active layer morphology and to increase the contact area between the active layer and the cathode, the complete device can be thermally treated after evaporation of the cathode (post-annealing) [1]. Overall, spin coating the active layer of a solar cell involves film formation via spreading of a fluid over a substrate, evaporation of the solvent from the fluid, various thermal treatments and evaporation of the cathode.

When spin coating, viscosity and surface tension must taken into account. However, different solutions can be spun regardless of small differences in viscosity and surface tension. This is not always the case with inkjet printing where minor changes in the two previously mentioned ink properties can hinder successful printing of a material. If the ink's surface tension is too high, instable droplets will be ejected from the print head which will generate inhomogeneous films. The viscosity of an ink influences the flow properties of the fluid in the print head. To increase long-term stability of the ink in the print head, high boiling point solvents are needed to prevent print head nozzle clogging [2-4]. Due to the incorporation of a high boiling component, printed films can be in a fluid state after printing. Therefore, an additional heating step is needed to dry the active layer which is not always the situation with spin coated films. After drying the active layer, the same processing steps are used for devices with printed layers as for devices with spin coated layers.

EXPERIMENT

Patterned indium tin oxide (ITO) substrates with a sheet resistance of roughly 20 $\Omega/\square$ were first cleaned with a detergent and then treated with O_2 plasma. For the HTL inkjet ink, PEDOT:PSS (AI 4083 from H.C. Stark) was combined with deionized water, ethylene glycol, isopropanol and TritonX-100. The active layer ink consisted of 0.33 wt% P3HT (Honeywell) in a 1:1 wt/wt combination with PCBM (Solenne, 99.5% purity) in chloro- and trichlorobenzene (55 wt%/45 wt%). After printing, HTL printed layers were dried in a stream of N_2 and annealed at 180°C for 15 minutes. Printed active layers were dried after printing at 100°C and pre-annealing was carried out at 100°C for 10 minutes. Printed HTL and active layers had thicknesses of 50 and 120 nm, respectively. For spin coating, PEDOT:PSS was diluted with isopropanol/ethanol/water (40 vol%/40 vol%/20 vol%) and spun at 3000 rpm. The active layer was spin coated from a solution with a concentration of 1.4 wt% P3HT (1:1 wt/wt with PCBM) in chlorobenzene at 1500 rpm. Spin coated HTL and active layers had thicknesses of 40 and 110 nm, respectively. After film preparation, aluminum (120 nm) was evaporated at 10^{-6} mbar which resulted in a device active area of 0.79 cm^2. Current-voltage characteristics of the solar cells were recorded under illumination with a Steuerangel solar simulator at 100 mW/cm^2 (AM 1.5) in conjunction with a Keithley 2400 source meter. The surface topographies of printed films were analyzed with atomic force microscope (AFM, Nanosurf easyScan) in non-contact and contact modes. Surface tension measurements were performed with a tensiometer (bubble pressure method) from SITA Messtechnik GmbH. The viscosities of the inkjet inks were measured with a cone and plate rheometer from Haake with a shear rate range from 0.1 to 1000 s^{-1}.

RESULTS AND DISCUSSION

During inkjet printing, droplets of an ink are ejected from a print head and these droplets must flow together to form a homogeneous film. The printer used for this experiment accomplishes this process by printing droplets which flow together to form lines and finally, the lines flow together to form a film. Figure 3 shows how the printer used for this report generated printed images.

Stages 1, 2 and 3 in Figure 3 are repeated until the desired resolution in the x and y axes of the image is reached. Overall, Figure 3 shows that the printing process involves many individual steps which take a non-trivial amount of time. When spin coating, film formation

155

usually can occur in less than 30 seconds. Printing can take up to several minutes for a given film depending upon the desired resolution, printing speed or solar cell area. Due to the length of time needed during inkjet printing, a combination of low and high boiling point solvents is used for inkjet solvent system to increase the stability of the ink in the print head.

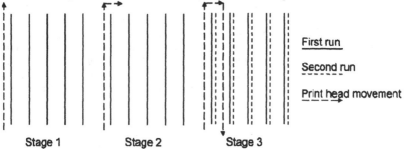

Figure 3. Three stages of printing where the first lines of the image are printed (stage 1), the print head moves perpendicular to the printed lines without printing (stage 2) and the next set of lines are printed as the print head moves back over the substrate (stage 3).

In this report, a solvent system consisting of chloro- and trichlorobenzene was utilized because it has been shown to completely dissolve P3HT and PCBM and to prevent nozzle clogging. We have demonstrated this solvent system where solar cells with inkjet printed passive and active layers were produced [8]. The surface tension and viscosity of a P3HT:PCBM ink consisting of chloro- and trichlorobenzene (55 wt%/45 wt%) was found to be 32.5 mN/m and 1.0 to 1.5 mPa*s, respectively.

In addition to the ink properties previously discussed, the surface topography of a printed film can influence solar cell performance. For the HTL, a smooth, homogeneous surface is desired because this layer acts as a 'bridge' between the anode and the active layer. As previously discussed, thermal treatments are needed for P3HT:PCBM thin films in order to induce the appropriate degree of phase separation between the two components. Ideally, a balance is needed between a high interfacial area between P3HT and PCBM and bigger, isolated domains of the two components in order to separate as many excitons are possible and to effectively transport free charges to the electrodes, respectively. Furthermore, it has been shown that various ratios of P3HT to PCBM and annealing temperatures can lead to different thin film surface topographies [9]. A useful method to analyze the surface topographies of printed films is AFM as shown in Figure 4. As a reference, the surface topographies of spin coated films from the same materials are shown.

The surface profiles of inkjet printed PEDOT:PSS shown in Figure 4 (top left) is rougher than the corresponding spin coated film where the roughness values were found to be 3 and 1 nm, respectively. Another difference for the two PEDOT:PSS films is the presence of pin holes in the film that was generated with inkjet printing. In contrast to the HTL films, the roughness values of the inkjet printed (bottom left) and spin coated (bottom right) P3HT:PCBM films are both close to 1 nm. However, the surface profile of the printed film shows small pin holes where the spin coated film is pin hole free. The film topography differences are proposed to be due to the printing process and/or the different solvents used for the respective inks.

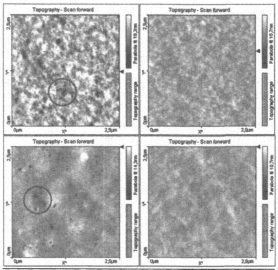

Figure 4. Surface topography images of inkjet printed (left) and spin coated (right) PEDOT:PSS (top) and P3HT:PCBM (bottom) films.

In order to further examine the differences between devices with inkjet printed and spin coated polymer layers, the current-voltage characteristics of the two types of devices were measured and plotted as shown in Figure 5.

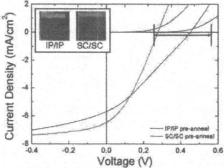

Figure 5. A plot of current density versus voltage for solar cells with spin coated (SC) and inkjet printed (IP) PEDOT:PSS and P3HT:PCBM layers after pre-annealing. The blue arrow indicates the range of V_{oc} values that are achievable upon post-annealing.

Despite the different film topographies shown in Figure 4, the η_{PCE} of solar cells with inkjet printed and spin coated HTL and active layers is 0.73 and 0.55%, respectively, after pre-annealing. The η_{PCE} values are different for both types of devices due to the variations in the V_{oc}. The blue arrow in Figure 5 shows the possible range of V_{oc} values for our devices with IP/IP or SC/SC layers which is achievable upon post-annealing at 150°C for 10 minutes with the previously described ink formulations. It was found that V_{oc} changes considerably for P3HT:PCBM solar cells with spin coated layers upon post-annealing where an increase from 0.20 to 0.55 V was found which resulted in η_{PCE} of greater than 2%. For devices with inkjet printed layers, lower η_{PCE} values after post-annealing were found due to a reduction in the V_{oc}. The cause for this effect is still under investigation. The η_{PCE} for both types of solar cells could be further increased upon incorporation of a LiF/Al cathode. In this work, an aluminum only cathode was used. The insert in Figure 5 shows a photograph of solar cells with inkjet printed (IP/IP) and spin coated (SC/SC) HTL and active layers. It is apparent from this image that printed devices suffer from film inhomogeneities. The edge effects for the device with printed layers could be due to the interactions of chloro- and trichlorobenzene upon drying due to the different boiling points of these two solvents (131 versus 213°C).

CONCLUSIONS

In this work, we have shown that the chloro-/trichlorobenzene solvent system has a suitable surface tension and viscosity to generate acceptable printed films for solar cell applications. Surface roughness values of roughly 3 nm were found for printed HTL and active layers. Finally, slightly better performance was found for devices with printed layers after pre-annealing and the η_{PCE} of both types of solar cells was strongly impacted by post-annealing at 150°C.

ACKNOWLEDGMENTS

The authors would like to thank Prof. Dr. Dieter Neher (University of Potsdam, Germany) for stimulating discussions and Steffi Kreissl (Fraunhofer IAP) for lab assistance. Funding was provided by the German Federal Ministry of Education and Research (BMBF), project 13N10317.

REFERENCES

1.   W. Ma, C. Yang, X. Gong, K. Lee and A.J. Heeger, *Adv. Funct. Mater.* **15**, 1617-1622 (2005).
2.   C.N. Hoth, S.A. Choulis, P. Schilinsky and C.J. Brabec, *Adv. Mater.* **19**, 3973-3978 (2007).
3.   C.N. Hoth, P. Schilinsky, S.A. Choulis and C.J. Brabec, *Nano Lett.* **8**, 2806-2813 (2008).
4.   T. Aernouts, T. Aleksandrov, C. Girotto, J. Genoe and J. Poormans, *Appl. Phys. Lett.* **92**, 033306 (2008).
5.   Y. Zhou, F. Zhang, k. Tvingstedt, S. Barrua, F. Li, W. Tian and O. Inganäs, *Appl. Phys Lett.* **146**, 233308 (2008).
6.   H. Schmidt, H. Flügge, T. Winkler, T. Bülow, T. Riedl and W. Kowalsky, *Appl. Phys. Lett.* **94**, 243302 (2009).
7.   F. Padinger, R.S. Rittberger, N.S. Sariciftci, *Adv. Funct. Mater.* **13**, 85-88 (2003).
8.   A. Lange, M. Wegener, C. Boeffel, B. Fischer, A. Wedel and D. Neher, Sol. Energy Mater. Sol. Cells **94**, 1816-1821 (2010).
9.   A. Swinnen, I. Haeldermans, M. vande Ven, J. D'Haen, G. Vanhoyland, S. Aresu, M. D'Olieslaeger and J. Manca, *Adv. Funct. Mater.* **16**, 760-765 (2006).

Charge Transport

Mater. Res. Soc. Symp. Proc. Vol. 1270 © 2010 Materials Research Society   1270-II08-07

High-power Organic Field-effect Transistors Using a Three-dimensional Structure

M. Uno[1,3], Y. Hirose[1,2], K. Nakayama[1], T. Uemura[1,2], Y. Nakazawa[1], K. Takimiya[4], and J. Takeya[1,2]

[1]Graduate School of Science, Osaka Univ., Toyonaka 560-0043, Japan. [2]ISIR, Osaka Univ., Ibaraki 567-0047, Japan. [3]TRI-Osaka, 2-7-1, Ayumino, Izumi, Osaka 594-1157, Japan. [4]Graduate School of Engineering, Hiroshima Univ. Higashi-Hiroshima 739-8527, Japan.

ABSTRACT

Three-dimensional organic field-effect transistors with multiple sub-micrometer channels are developed to exhibit high current density and high switching speed. The sub-micrometer channels are arranged perpendicularly to substrates and are defined by the height of a multi-columnar structure fabricated without using electron-beam-lithography technique. For devices with dinaphtho[2,3-b:2',3'-f]thieno[3,2-b]thiophene, extremely high current density exceeding 10 A/cm^2 and fast switching within 200 ns are realized with an on-off ratio of 10^5. The unprecedented performance is beyond general requirements to control organic light-emitting diodes, so that even more extensive applications to higher-speed active-matrices and display-driving circuits can be realized with organic semiconductors.

INTRODUCTION

Progress in organic electronics have been remarkable in these decades because of their attractiveness such as compatibility of environment-friendly and low-cost production processes, flexibility, light weight and resistivity against mechanical shock. In developing state-of-the-art organic thin-film transistors (OTFTs), as well as a material approach to realize high-mobility π-conjugated organic compounds [1-4], the improved performances are also attributed to advanced techniques such as use of ultrathin dielectrics [5], and crystallization technique of organic semiconductors [6, 7]. Furthermore, we recently proposed to exploit availability of the three-dimensional space, constructing multiple vertical channels densely packed in a limited area [8, 9]. The three-dimensional OFETs (3D-OFETs) indeed showed outstanding performance with output current up to 0.6 A/cm^2.

In this work, we have developed a sub-micrometer-channel version of the 3D-OFETs by optimizing the etching process using a reactive gas mixture to fabricate highly-perpendicular side-walls, without any low-throughput processes such as electron-beam lithography. An advantage to adopt such short-channels is that high-speed response can be expected for current gain with the application of gate voltage V_G. The other advantage is that output drain current I_D is further gained with shorter-channel devices because I_D in the linear regime is inversely proportional to channel length L. In other words, the structure of the short-channel 3D-OFETs compensates for small μ, which is often regarded as a general disadvantage of organic semiconductors. We present here the sub-micrometer 3D-OFETs with very high output current exceeding 10 A/cm^2, very high switching speed within 0.2 μs and fair on-off ratio of 10^5, by reproducing more ideal vertical metal-insulator-semiconductor (MIS) structure and by preserving good molecular order.

EXPERIMENT

Figure 1 shows schematic structures of a planar OFET and the 3D-OFET. In the 3D-OFET, semiconductor channels are built on the sidewalls of the 3D structure, so that injected carriers flow vertically from the source electrode on the substrate to the drain electrode on the land part of the structure. The devices are fabricated using similar processes as reported previously [9], as shown in Fig. 2, except in the process of etching Si substrates. The 3D structure is fabricated by reactive ion etching of the Si substrate using mixed gases of SF_6 and CHF_3 with the low rate of 3.0 sccm and 0.8 sccm, respectively, at the total gas pressure of 27.0 Pa.

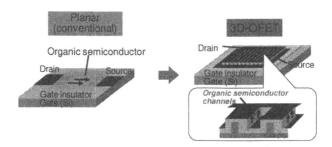

Figure 1. Schematic illustration of a planar OFET and the 3D-OFET.

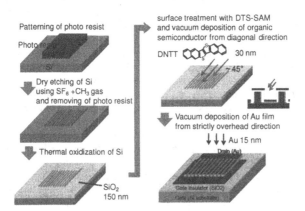

Figure 2. Fabrication processes for 3D-OFETs. Molecular structure of DNTT is also shown.

162

Mixing of the CHF₃ gas has an effect of forming a sidewall protection layer, which enables fabrication of the highly perpendicular sidewalls. In order to characterize the effect of channel length, L is varied from 0.5 μm to 2.5 μm by simply changing the etching time. After the dry-etching process, Si is thermally oxidized to the thickness of 150 nm to make a SiO_2 gate-insulating layer. As the previous experiments, a self-assembled monolayer of alkyl-silanes is coated and DNTT is deposited with the rate of 0.08 nm/s from a diagonal direction of about 45 degree. Finally, gold is evaporated from the direction strictly perpendicular to the substrates to form both source and drain electrodes simultaneously. The size of a 3D-OFET is 100 μm x 100 μm and the pitch between adjacent vertical walls is 5 μm. As the result, the channel width is as long as 1100 μm even when one side of the sidewalls are coated with organic semiconductors, so that the channel ratio W/L is calculated to be as high as 2200 with the channel length of 0.50 μm. Figure 2(a) shows a scanning electron microscope (SEM) image of one 3D-OFET, and Figs. 2(b) and 2(c) show expanded views of their refined structures with the channel lengths of 0.9 μm. Indeed, the channels are formed in highly perpendicular direction to the substrate, and the gold films deposited on the land part and in groove part of the structures are clearly separated with each other. The channel lengths are measured in the real images for all the devices. The channel widths are estimated from the whole sidewalls covered with the organic semiconductors. FET characteristics of the devices were measured in ambient condition using Agilent B1500A semiconductor parameter analyzer for static transfer and output characteristics, and B1530A waveform generator/fast measurement units were used for the dynamic measurements of their response time.

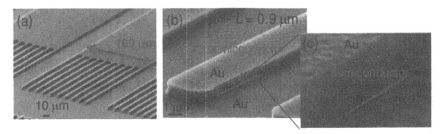

Figure 3. (a) A SEM image of the 3D-OFETs, and (b) (c) expanded views of the channels with the length of 0.9 μm.

RESULTS AND DISCUSSION

High output current of 3D-OFETs

Figures 4 shows typical transfer characteristics of the present 3D-OFETs with the channel lengths of 0.5 μm, 0.9 μm, and 2.5 μm, respectively. Drain voltage is set to -5 V in the measurements for all the devices. Remarkably high output current density of 12.6 A/cm² is obtained for the shortest device with V_G of -20 V, while maintaining fair on-off ratios of above 10⁵. The values of carrier mobility in the films formed on the sidewalls are estimated to be around 0.3 cm^2/Vs for the all devices. Although the values of the mobility are still less than those with lateral devices [4], it is suggested that good π-packed ordering of the molecules is preserved on the vertical sidewalls of the multi-columnar 3D structures. Figures 5(a)-(c) show output characteristics of the same devices. In the devices with the longer channel lengths, the drain currents exhibit a saturation trend in low V_G region, while the saturation is suppressed with the decrease of the channel length, due to the deviation from the gradual channel approximation with the gate dielectric thickness comparable to the channel length. Also, the drain currents exhibit nonlinear characteristics at low drain in Figs. 5(a) and 5(b) voltages because the ratio of the contact resistances on the total channel conductances becomes larger, which is common to short-channel devices. We are currently improving the process to minimize the dielectric thickness and to reduce the contact resistances.

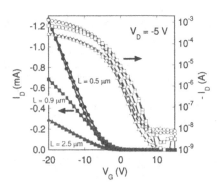

Figure 4. Transfer characteristics of the 3D-OFETs with channel lengths of 0.50 μm(blue filled circle), 0.90 μm(green filled square), and 2.5 μm(red filled triangle), respectively. Log(I_D) vs V_G plots are shown on the right axis with the same open symbols.

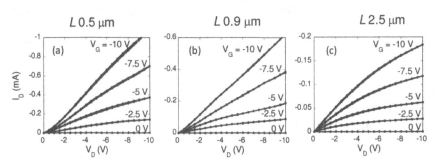

Figure 5. Output characteristics of the 3D-OFETs with channel lengths of (a) 0.50 μm, (b) 0.90 μm, and (c) 2.5 μm.

164

Transient response of the submicrometer-channel 3D-OFETs

We have examined dynamic response of the 3D-OFETs with 0.5-μm channel length to demonstrate the fast switching of the short-channel device. Pulsed gate voltages of 500 kHz are repeatedly applied with the height of -10 V, as shown in Fig. 6(a), while V_D is kept at -5 V. The rise time and the fall time of the input pulses are both set to 10 ns, which is actually the resolution limit of the equipment. Figure 6(b) shows the transient drain current in response to the pulsed V_G applied. The time to reach the on state is typically less than 0.2 μs; I_D reaches to ninety percent of the maximum value in 160 ns after V_G is applied, which corresponds to the response frequency of 6.3 MHz. The observed spike-like discharging current indicates significant influence of the parasitic capacitance in our devices. Though the present 0.5-μm-channel 3D-OFET includes significant parasitic capacitance about 20 times larger than the channel capacitance, the presently obtained response frequency is among the highest for organic transistors reported in the literature [12-14]. It is to be emphasized that the submicron-channel devices, which exhibit the very high switching speed and extremely high output current density, are fabricated by using only mass-producible photolithography technique. With the performance exceeding general requirement in matrix-controlling elements for OLED displays, the 3D-OFETs can open the way to the development of display-driving integrated logic circuits with organic semiconductors. The extremely high density of the output current itself would be useful even for other application such as power devices combined with their capability of their fast switching operation.

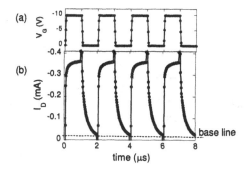

Figure 6. (a) Pulsed gate voltage of 500 kHz applied with the height of -10 V, and (b) transient response of the drain current of 3D-OFETs in response to the gate voltage. V_D is kept at -5.0 V during the measurement.

CONCLUSIONS

We have developed three-dimensional organic field-effect transistors with sub-micrometer channel length by photolithography and simple etching processes, establishing a technique of fabricating highly perpendicular vertical sidewalls. The device with 0.5-micrometer channel length demonstrated unprecedentedly high output current density of 12.6 A/cm^2 and high on-off ratio of 10^5, even with moderate carrier mobility about 0.3 cm^2/Vs. With the benefit of the short channel lengths, the 3D-OFETs show very high switch-on time within 0.2 μs, still leaving possibility to be reduced by more than one order of magnitude. The above results demonstrate that the 3D-OFET structure gives a practical compensation for relatively low values of carrier mobility which has been regarded as the weak point of organic semiconductor transistors.

ACKNOWLEDGMENTS

We gratefully thank to Nihon Kayaku Co., ltd for supplying DNTT. This work was financially supported by Industrial Technology Research Grant Program from NEDO, Japan, and by a Grant-in-Aid for Scientific Research (Nos. 21108514(π-space) and 22245032) from MEXT, Japan.

REFERENCES

1. Y. Y. Lin, D. J. Gundlach, S. F. Nelson, and T. N. Jackson, *IEEE Electron Device Lett.* **18**,606 (1997).
2. J. E. Anthony, J. S. Brooks, D. L. Eaton, and S. R. Parkin, *J. Am. Chem. Soc.* **123**, 9482 (2001).
3. H. Yan, Y. Zheng, R. Blache, C. Newman, S. Lu, J. Woerle, and A. Facchetti, *Adv. Mater.* **20**, 3393 (2008).
4. T. Yamamoto, K. Takimiya, *J. Am. Chem. Soc.* **129**, 2224 (2007).
5. H. Klauk, U. Zschieschang, J. Pflaum, and M. Halik *Nature* **445**, 745 (2007).
6. K. C. Dickey, J. E. Anthony, and Y.-L. Loo, *Adv. Mater.* **18**, 1721 (2006).
7. T. Uemura, Y. Hirose, M. Uno, K. Takimiya, and J. Takeya, *Appl. Phys. Exp.* **2**, 111501 (2009).
8. M. Uno, Y. Tominari, J. Takeya, *Appl. Phys. Lett.* **93**, 173301 (2008).
9. M. Uno, I. Doi, K. Takimiya, and J. Takeya, *Appl. Phys. Lett.* **94**, 103307 (2009).
10. N. Stutzmann, R. H. Friend, H. Sirringhaus, *Science* **299**, 1881 (2003).
11. K. Fujimoto, T. Hiroi, K. Kudo, and M. Nakamura, *Adv. Mater.* **19**, 525 (2007).
12. M. Kitamura and Y. Arakawa, *Appl. Phys. Lett.* **89**, 223525 (2006).
13. Y. Y. Noh, N. Zhao, M. Caironi, and H. Sirringhaus, *Nature Mater.* **2**, 784 (2007).
14. V. Wagner, P. WÄobkenberg, A. Hoppe, and J. Seekamp, *Appl. Phys. Lett.* **89**, 243515 (2006).

Mater. Res. Soc. Symp. Proc. Vol. 1270 © 2010 Materials Research Society   1270-II08-09

Monolithic Complementary Inverters Based on Intrinsic Semiconductors of Organic Single Crystals

T. Uemura[1,2], M. Yamagishi[1], Y. Okada[1], K. Nakayama[1], M. Yoshizumi[1], M. Uno[1], Y. Nakazawa[1], and J. Takeya[1,2]

[1]Graduate School of Science, Osaka University, Toyonaka 560-0043, Japan.
[2]ISIR, Osaka University, Ibaraki 567-0047, Japan.

ABSTRACT

A novel monolithic complementary inverter is fabricated on a single platform of an organic semiconductor crystal only by patterning low- and high-work-function metals to inject electrons and holes separately. With the benefit of using high-performance organic single-crystal transistors, the inverter indeed shows excellent performances with very low power consumption, high output gain, large noise margin, and small hysteresis.

INTRODUCTION

In the history of the silicon semiconductor technology, the invention of integrated circuits (ICs) has yielded tremendous profit in the practical economy. The integration technology has been based on fabricating numerous device components on a same monolithic *single-crystal* platform, allocating sets of *p*- and *n*-channel metal-oxide-semiconductor field-effect transistors (MOSFETs) for complementary NOT-logic elements, for example. Recently, organic semiconductors are argued as an emerging candidate for next generation semiconductor materials because of their unique advantages in low-cost fabrication processes near room temperature, light weight and mechanical flexibility. So far, however, their compatibility to the monolithic circuitry integration is not yet examined on a single-component platform. In this paper, we present a novel monolithic complementary inverter on a single-crystal organic semiconductor in order to present viability in developing organic integrated circuits.

This study is strongly motivated by remarkable progress in material development, device fabrication techniques and fundamental understanding of the charge transport mechanism, resulting in much improved performances of recent organic field-effect transistors (OFETs). The intrinsic carrier transport mechanism is proven to be band transport in high-mobility organic single-crystal transistors [1-3], as in silicon single crystals used for the ICs. Already several groups reported that mobility of organic single crystals is as high as 20 cm^2/Vs or even higher [4-6], so that 100 MHz clock frequency can be predicted with micrometer-channel devices assuming a standard charging model. Furthermore, techniques of solution-based crystallization is being developed very recently [7-9], so that even mobility as high as 5 cm^2/Vs is achieved from solution [10], suggesting that low-cost production routes on flexible panels can serve for mass-producing the platforms of low-price and medium-performance integrated circuits.

At present, fairly good inverter performances have been reported only for those with two different materials of *p*- and *n*-channel organic-semiconductor compounds [11-14]. In this configuration, one has to concern about fabrication process to distribute the two organic materials in a miniaturized space for integration. Considering a material cost and circuit downsizing for low-power and high-speed operation, fabrication of circuits on a single-component platform is favorable, which realized in the silicon technology where p- and n-MOSFETs are designed on the same single-crystal platform using precise photolithography and doping techniques. Based on this viewpoint, one-component complementary inverters are also developed using ambipolar organic semiconductors with narrow energy gaps; however, they inherently suffer from high off-current with relatively large number of thermally excited carriers due to the narrow gaps themselves, which is harmful to preserve the merit of the CMOS architecture, i.e. low power consumption upon the logic operation [15]. Their reported mobility, which is essential for high-speed CMOS operation, is not as high as common p-type devices, either [16].

We use wider-gap rubrene single-crystal semiconductors because of their much better on-off ratio as used for OFETs. Rubrene crystal transistors are usually p-type combined with the source electrode of a noble metal such as gold, however, they can be n-type if a low-work-function metal such as calcium are used for the electrode [17]. Our present monolithic CMOS is fabricated on the rubrene single crystal using both gold and calcium electrodes to inject holes and electrons, respectively. At present, we measured the characteristics in an inert atmosphere. However, we note that passivation films with inorganic layers are effective for their air-stable operation, so that the technique can be used to bring out the monolithic organic CMOS in atmosphere in principle. The present design is grounded on the idea that such organic semiconductors as rubrene are intrinsic semiconductors, contrastingly to the silicon single-crystal CMOS, where both n-type and p-type layers are selectively fabricated by ion-injection and the operation is due to the formation of the inversion layers. Therefore, the organic-crystal CMOS can be fabricated with reduced processes without the ion-injection. In addition to the reduced fabrication steps, the devices hold general merits of monolithic design such as facilitated circuitry integration and accelerated clock frequency without unnecessary transmission delays. In the following, we show performances of the rubrene CMOS inverter together with the transistor characteristics of the two unipolar channels on the same crystal. Because of high mobility and high on-off ratio of the rubrene single-crystal transistors, the inverter show excellent performances of low power consumption, high gain and small hysteresis upon switching.

EXPERIMENT

The schematic configuration of the monolithic complementary inverter is illustrated in Fig. 1(a) and an optical image of a typical device is shown in Fig. 1(b). In the present monolithic CMOS, both p- and n-channel are fabricated on the same rubrene single crystal by simply patterning the electrodes with gold and calcium to inject holes and electrons, respectively. First, two gate electrodes of Cr/Au were separately patterned for the p- and n-channels on a doped-Si/SiO$_2$ (500 nm) substrate. For the main gating, 200-nm-thick poly(methyl methacrylate) (PMMA) was spin-coated on the surface in air and annealed at argon atmosphere. The proper choice of the polymer dielectric was necessary to minimize the electron traps on the surface of the dielectric layer, as reported in literatures [18]. The rubrene crystal was then laminated on the PMMA surface by an electrostatic force [19]. The top-contact electrodes of Ca and Au were fabricated on the single crystal through metal masks. The dimensions of the two channels were adjusted for a balanced inverter operation considering the differences of mobility and threshold voltages on both channels. After the deposition of the electrodes, the devices were kept in Ar-filled grove-box (O_2, H_2O < 0.1 ppm) and the whole electrical measurements were performed there.

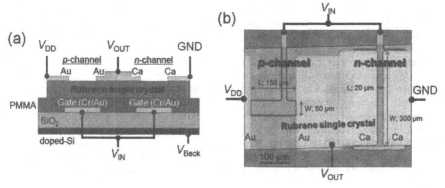

Figure 1. (a) Schematic illustration of a monolithic complementary inverter using a rubrene single crystal. (b) Optical view of the typical device. The p- and n-channel were separately fabricated on the same rubrene crystal. The channel length (L) and width (W) of p-channel were 150 and 50 μm and n-channel were 20 and 300 μm, respectively.

As shown in Fig. 1(b), the two separated gold gate electrodes are connected outside the devices and both are used for the V_{IN} terminal for the inverter operation. In the middle of the p- and n-channels, Ca and Au electrodes are overlapped to make the V_{OUT} terminal. The left Au electrode and the right Ca electrode are used for V_{DD} and GND terminals, respectively. We note that the bottom layers of SiO_2 and doped Si are essential to avoid cross talking between the p- and n-channels by applying "back-gate" voltage V_{Back}. Though the p- and n-channels are apart with each other by 100 μm, we experienced that minority carriers can diffuse into the other channel, so that the n-channel transistor can show ambipolar function without the V_{Back} application. Since it causes current leakage between the two transistors, resulting in additional power consumption, it is necessary to electrically separate the devices throughout the inverter operation. With the application of V_{Back} from the doped-Si, unnecessary carriers are removed by depleting the semiconductor outside the two transistors.

Figure 2 shows the best performances of the two transistors that compose the monolithic rubrene crystal inverter. Figures 2(a) and 2(b) are transfer characteristics for the p- and n-channel transistors in a saturation regime, respectively, while Figs. 2(c) and 2(d) are their output characteristics. The gray curves represent the characteristics without the application of V_{Back}. In Fig. 2, only n-channel transistor shows ambipolar operation due to hole diffusion from p-channel to n-channel region. In order to suppress the hole diffusion, we applied +10 V to V_{Back} terminal. As a result of V_{Back} of +10 V, the two transistors are well separated with each other, showing typical unipolar characteristics with well-defined off states. Since the electron diffusion to p-channel region could not be observed, the back-gate voltage did not affect the transfer characteristics of p-channel. The different properties of carrier diffusion are assumed to be caused by the intrinsic properties of carrier diffusion in rubrene crystal. The hole and electron mobility are estimated to be 1.2 and 0.2 cm^2/V, respectively, representing standard values for the top-contact rubrene single-crystal transistors [17]. We note that the mobility is not as large as typical bottom-contact devices because of possible damage in the crystal during the evaporation of the metals and of additional resistances in the direction of the thickness. The subthreshold swing S is as small as 0.6 V/decade and 0.5 V/decade for p- and n-channel, respectively, indicating that the interface of the rubrene crystal to the PMMA is formed with small density of trap states. Very small hysteresis and small threshold voltages, which are -1 V and +5 V for p- and n-channel devices, respectively, are also due to clean interface with the organic single crystal. Both channels are operated within ±10 V, which is adequately small for a realistic power source for a logic circuit in practical applications. We note that the output characteristics show good saturation behavior without indication of injection barrier even for the n-channel transistor. This is because the energy difference between the work function of Ca (~2.9 eV) and the lowest unoccupied molecular orbital (LUMO) level of rubrene (~3.2 eV) is adequately small for the electron injection. The small energy difference is almost identical to that between the work function of Au (~5.1 eV) and the highest occupied molecular orbital (HOMO) level (~5.4 eV). The sufficiently large HOMO-LUMO gap of rubrene is helpful to preserve reasonable on/off ratio in both channels, leading to low-power consumption in the inverter operation.

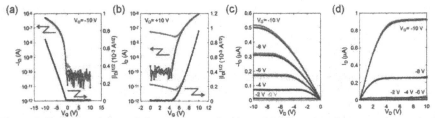

Figure 2. Transistor characteristics of the p- and n-channel fabricated on the same rubrene single crystal. (a) Transfer characteristics of p-channel and (b) n-channel with applying a back gate voltage of +10 V. (c) Output characteristics of p-channel and (d) n-channel in the same condition. In each plot, the characteristics without applying a back-gate voltage are shown in gray plots. Considering their inverter operation of each channel, in these measurements, gate and drain voltages were applied to common V_{IN} and V_{OUT} terminals, respectively. The other terminals of V_{DD} and GND terminals were connected to the ground terminal.

Figure 3 shows the characteristics of the inverter switching with the application of the back-gate voltage of +10 V. The characteristics without the back-gating are plotted in gray. In Fig. 3(a), voltage transfer characteristics with the supply voltage V_{DD} of +8 V, +9 V and +10 V are presented, exhibiting that the monolithic crystal inverter indeed operates properly with small hysteresis. The switching threshold voltages at the crossing points to the line of $V_{OUT}=V_{IN}$ were 5.8 V, 6.1 V and 6.4 V when $V_{DD}=$+8, +9 and +10 V, respectively. These values are close to the ideal values calculated a $0.5V_{DD}$, meaning that sufficient noise margin is secured for the logic operation in this device. Figure 3(b) shows the DC gain defined by dV_{OUT}/dV_{IN} reaching as high as 140 at $V_{DD}=$+10 V, which is comparable to the values reported for high-performance complementary inverters composed of organic semiconductors [11-14]. The result of the high DC gain is attributed to the high switching performances (small S values) in both p- and n-channel transistors. Figure 3(c) shows the penetration current I_{DD} between the V_{DD} and GND terminals during the inverter switching. Since the product of I_{DD} and V_{DD} represents the power consumption in a logic inversion, the result indicates that the power consumption in the switching state is in the order of µW, while it is in the order of nW in the static states when both transistors are in the off state, which is again plausible among the high-performance organic CMOS devices [11-14]. The application of V_{Back} is essential again to preserve the low power consumption, which is regarded as a general benefit of complimentary inverters.

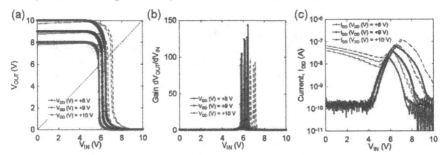

Figure 3. The switching characteristics of a monolithic complimentary inverter. (a) Voltage transfer characteristics of a monolithic complementary inverter with applying a back gate voltage of +10 V. (b) Corresponding DC values of inverter switching. (c) Penetration current between the V_{DD} and GND terminals at the switching. In each plot, the characteristics without applying a back gate voltage are shown in gray plots.

Figure 4 demonstrates switching performance of the inverter repeated at 1 Hz with the application of V_{Back}; the square-wave signal input to the V_{IN} terminal shown in Fig.4 (a) is correctly inverted at the output terminal shown in Fig.4 (b). Since the inverter consumes only a few nW in the static state, the present CMOS inverter indeed realizes the logic inversion with minimum power consumption, demonstrating the well-known advantage of the complementary circuitry.

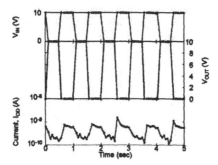

Figure 4. Demonstration of repeat switching of a monolithic complementary inverter with applying a back gate voltage of +10 V. (a) Square waves of input voltages to V_{IN} terminal. (b) Output voltages detected at V_{OUT} terminal. (c) Penetration current through the complementary inverter.

The comparison between the device performances with and without the V_{Back} application demonstrates that electrical separation of the p- and n-channel transistors is essential in proper operation of the organic CMOS inverter. Without V_{Back}, the n-channel transistor shows an ambipolar operation due to the parasitic carrier diffusion from the p-channel, which significantly reduces its performances; the S value increases from 0.5 V/decade to 3 V/decade and the on/off ratio is reduced from 10^4 to 10^1. As the result, the inverter switching performances such as power consumption, gain, noise margin, and hysteresis are all deteriorated. The most serious problem appears in the off-state power consumption close to μW, which is fatal in repeated logic operations. We note that the back-gating technique is not useful in the ambipolar devices based on narrow-gap organic semiconductors with the same source-electrode material for the two channels, because it does not reduce the injection of minority carriers. Though it might be easier to fabricate monolithic inverters with the narrow-gap semiconductors which operates in air, ambipolar devices have been shown to have significant disadvantage in power consumption. On the other hand, the present monolithic device possesses excellent performance but it does not operate in air at present. The development of passivation techniques using both organic and inorganic layers are under way in the authors' group, so that the monolithic rubrene inverter can be used in air. Finding possible combination of other organic semiconductor crystals and air-stable metals can give other solutions.

CONCLUSIONS

In conclusion, we demonstrate the operation of a novel monolithic complementary inverter using an organic single-crystal semiconductor. This monolithic CMOS can be fabricated on the same organic monolithic substrate only by patterning low- and high-work-function metals to inject electrons and holes separately. The demonstration represents the unique nature of the organic semiconductor to be an intrinsic semiconductor, consisting of only one molecular component. With the benefit of using high-performance organic single-crystal transistors, the inverter indeed shows excellent performances with very low power consumption, high output gain, large noise margin, and small hysteresis. Furthermore, the advantage of minimized fabrication steps in the monolithic design helps the circuitry integration without unnecessary transmission delays, so that future applications to low-cost RF-ID tags, display driver circuits, and logic devices in wearable electronics can be benefited.

ACKNOWLEDGMENTS
This work was financially supported by Industrial Technology Research Grant Program in 2006 and 2009 from NEDO, Japan, and by a Grant-in-Aid for Scientific Research (Nos. 17069003, 19360009, and 21108514) from MEXT, Japan.

REFERENCES
1. J. Takeya, K. Tsukagoshi, Y. Aoyagi, T. Takenobu, and Y. Iwasa, *Jpn. J. Appl. Phys.* **44**, L1393 (2005).
2. V. Podzorov, E. Menard, J. A. Rogers, and M. E. Gershenson, *Phys. Rev. Lett.* **95**, 226601 (2005).
3. J. Takeya, J. Kato, K. Hara, M. Yamagishi, R. Hirahara, K. Yamada, Y. Nakazawa, S. Ikehata,

K. Tsukagoshi, Y. Aoyagi, T. Takenobu, and Y. Iwasa, *Phys. Rev. Lett* **98**, 196804 (2007).

4. E. Menard, V. Podzorov, S.-H. Hur, A. Gaur, M. E. Gershenson, and J. A. Rogers, *Adv. Mater.* **16**, 2097 (2004).

5. O. D. Jurchescu, M. Popinciuc, B. J. van Wees, and T. T. M. Palstra, *Adv. Mater.* **19**, 688 (2004).

6. J. Takeya, M. Yamagishi, Y. Tominari, R. Hirahara, Y. Nakazawa, T. Nishikawa, T. Kawase, and T. Shimoda, *Appl. Phys. Lett.* **90**, 102120 (2007).

7. K. C. Dickey, J. E. Anthony, and Y.-L. Loo, *Adv. Mater.* **18**, 1721 (2006).

8. T. Minari, M. Kano, T. Miyadera, S.-D. Wang, Y. Aoyagi, and K. Tsukagoshi, *Appl. Phys. Lett.* **94**, 093307 (2009).

9. S. S. Lee, C. S. Kim, E. D. Gomez, B. Purushothaman, M. F. Toney, C. Wang, A. Hexemer, J. E. Anthony, and Y.-L. Loo, *Adv. Mater.* **21**, 3605 (2009).

10. T. Uemura, Y. Hirose, M. Uno, K. Takimiya, and J. Takeya, *Appl. Phys. Exp.* **2**, 111501 (2009).

11. H. Klauk, U. Zschieschang, J. Paum, and M. Halik, *Nature* **445**, 745 (2007).

12. M. P. Walser, W. L. Kalb, T. Mathis, T. J. Brenner, and B. Batlogg, *Appl. Phys. Lett.* **94**, 053303 (2009).

13. M. P. Walser, W. L. Kalb, T. Mathis, and B. Batlogg, *Appl. Phys. Lett.* **95**, 233301 (2009).

14. X.-H Zhang, W. J. Potscavage, S. Choi, and B. Kippelen, *Appl. Phys. Lett.* **94**, 043312 (2009).

15. T. D. Anthopoulos, S. Setayesh, E. Smits, M. Cölle, E. Cantatore, B. de Boer, P. W. M. Blom, and D. M. de Leeuw, *Adv. Mater.* **18**, 1900 (2006).

16. R. W. I. de Boer, A. F. Stassen, M. F. Craciun, C. L. Mulder, A. Molinari, S. Rogge, and A. F. Morpurgo, *Appl. Phys. Lett.* **86**, 262109 (2005).

17. T. Takahashi, T. Takenobu, J. Takeya, and Y. Iwasa, *Appl. Phys. Lett.* **88**, 033505 (2006).

18. L.-L. Chua, J. Zaumseil, J.-F. Chang, E. C.-W. Ou, P. K.-H. Ho, H. Sirringhaus, and R. H. Friend, *Nature* **434**, 194 (2005).

19. J. Takeya, C. Goldmann, S. Haas, K. P. Pernstich, B. Ketterer, and B. Batlogg, *J. Appl. Phys.* **94**, 5800 (2003).

**Poster Session: Organic Materials
for OFETs, OLEDs, and OPV**

Mater. Res. Soc. Symp. Proc. Vol. 1270 © 2010 Materials Research Society   1270-II09-02

Copolymers Based on 2,7-Carbazole With Alkylated and Perfluoroalkylated Thiophene Derivatives.

Pierre-Luc T. Boudreault[1], Badrou Reda Aïch[1,2], Anne-Catherine Breton[1], Ye Tao[2], Mario Leclerc[1]

1) Canada Research Chair on Electroactive and Photoactive Polymers, Département de Chimie, Université Laval, Quebec City, Quebec, Canada, G1K 7P4
email : mario.leclerc@chm.ulaval.ca
2) Institute of Microstructural Sciences, National Research Council of Canada, Ottawa, Ontario, Canada, K1A 0R6

ABSTRACT
We report the synthesis and characterization of new copolymers based on 2,7-carbazole and new thiophene-based comonomers bearing either regular alkyl or perfluoroalkylated side chains. The characterization of these new polymers helped us to study the effect of the fluorine atoms on the self-assembly of the resulting copolymers in the solid state. XRD and DSC results have shown interesting results regarding the packing and organization of the materials. Finally, OFETs were fabricated to confirm or refute the promising properties we had observed.

INTRODUCTION
The synthesis of new air-stable polymers for organic field-effect transistors (OFETs) has attracted a lot of interest. Recent performances are getting closer to those observed with small molecules. More importantly they are close to the performances of amorphous silicon (α-Si) which is the main target. Their properties such as light weight, flexibility, low temperature processing, and facile tuning of the optical and electrical properties, make them suitable to replace α-Si in low cost and disposable devices.
Besides poly(3-hexylthiophene),[1] only a few polymers have reached mobility above 0.1 cm^2/(V.s) by solution processing. Among them, Müllen et al. have recently been able to obtain hole mobility above 1 cm^2/(V.s) with a donor-acceptor copolymer.[2] However, this polymer is rather unstable under ambient conditions and needs to be tested under a controlled atmosphere to reach these high performances. One of the most promising polymers was developed by Ong et al. where they were able to obtain a mobility of 0.25 cm^2/(V.s) under ambient conditions.[3] McCullough et al. have shown that it is possible even with highly disordered polymers to obtain good mobilities that are highly reproducible.[4, 5] With this type of polymer, hole mobilities as high as 0.13 cm^2/(V.s) were observed. Some 2,7-disubstituted carbazole[6] have recently shown excellent thermal and air stability of the device with a hole mobility of 10^{-2} cm^2/(V.s).[7]
Along these lines, we will present the synthesis of completely new copolymers based on carbazole and thiophene units with regular or perfluorinated alkyl chains to force the self assembly which could improve the charge transport properties.

EXPERIMENTAL DETAILS
Synthesis

The alkylated thiophene (bithiophenes and terthiophene) derivatives were synthesized as reported in the literature[4] and the dialkylated quaterthiophene **4** was obtained from the same method and the synthesis is shown in Figure 1. [8] On the other hand, the perfluoroalkylated thiophenes were obtained from a completely different pathway as described by Marks et al.[9, 10]. In order to obtain the 3-bromo-2,2'-bithiophene, a Grignard reaction was performed between 2-bromothiophene and 2,3-dibromothiophene. The synthesis of both comonomers is straightforward; the perfluorinated side chains are added to the 3-bromo-2,2'-bithiophene and the 2,2':5',2''-terthiophene through copper coupling. In both cases, the final step is the bromination of the terminal positions of the thiophene derivatives to complete the synthesis of the comonomers.

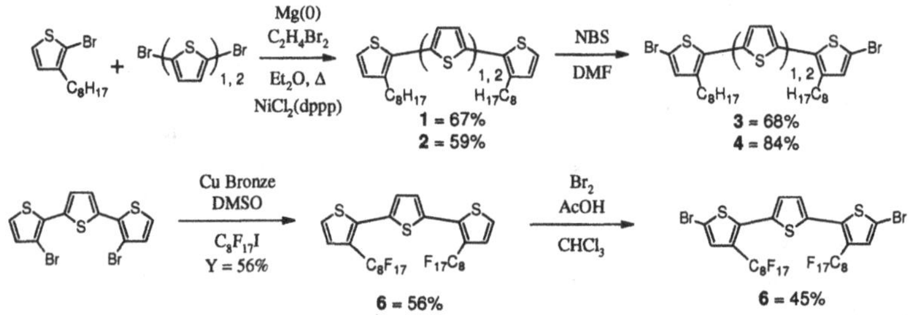

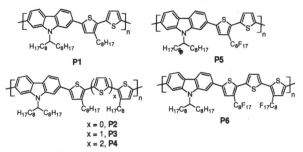

Figure 1. Synthesis of the monomers based on thiophene bearing regular and perfluorinated side chains.

The other comonomer is the 2,7-Bis(4',4',5',5'-tetramethyl-1',3',2'-dioxaborolan-2'-yl)-N-9''-heptadecanylcarbazole which has been used thoroughly during the past few years in order to obtain high molecular weight poly(2,7-carbazole) derivatives.[11] It was first thought that only linear octyl side chains would be used on the carbazole moiety. However, after many polymerization reactions, the polymers obtained had very low solubility and molecular weights below 5.0 kDa. Copolymers **P1** to **P6** were all synthesized with Suzuki cross-coupling polymerization, the chemical structure are shown in Figure 2. The conditions were slightly different from a regular Suzuki cross-coupling. The catalyst was Pd$_2$dba$_3$, the ligand allowing the higher molecular weight was P(o-tol)$_3$ while the base and solvent were tetraethylammonium hydroxide and toluene respectively.

Figure 2. Chemical structure of new alternating copolymers based on alkylated and perfluorinated thiophenes.

After washing the polymer with acetone and hexanes with a Sohxlet apparatus, the polymers were dissolved in CHCl₃ and precipitated in methanol. The polymerization yields were acceptable to excellent (**P1**=78%; **P2**= 59%; **P3**=97%; **P4**=75%; **P5**=84%; **P6**= 83%).

Polymer Characterization

After purification of the polymers, they were characterized using common techniques to obtain their molecular weights, thermal, optical and electrochemical properties. All these results are presented in Table 1. For the regular alkyl chains, the molecular weights were all very good except for **P2** (M_n = 8.5 kDa). This is probably due to some impurities left in the 2,2'-dibromo-4,4'-di-*n*-octylbithiophene. This compound is a viscous oil and consequently, could only be purified by column chromatography. On the other hand, the copolymers containing perfluoroalkylated side chains have allowed M_n of 11.7 and 18.9 kDa for **P5** and **P6**, respectively. Even if the M_n are relatively different, the degree of polymerization is still around 10 for both copolymers. We suppose that when 10 repeating units have been reached the polymers are not soluble in the reaction solvent anymore, it then precipitates and the polymerization reaction stops.

Table 1. Optical absorption maxima (λ_{max}) and electrochemical properties of the new materials

Oligomer	$M_n : M_w$ [kDa]	T_d [°C]	T_g [°C]	λ_m [nm]	E_g^{opt} [eV] Thin film [a]	E_{HOMO} [eV] [b]
P1	19.5: 41.3	470	-	432	2.51	- 5.61
P2	8.5: 15.4	-	-	402	2.74	- 5.69
P3	51: 110	440	-	432	2.46	- 5.47
P4	23.1: 49.0	470	154	445	2.40	- 5.39
P5	11.7: 25.4	320	95	391	2.76	- 5.34
P6	18.9: 32.8	370	80	408	2.65	- 5.81

[a] Measurements performed on a spin-coated film on quartz. [b] Ionization potential (E_{HOMO}; HOMO: highest occupied molecular orbital) measured from the cathodic onset.

The thermal properties have shown that the polymers containing perfluorinated chains are less thermally stable than their regular alkyl chains counterparts. Both copolymers with perfluorinated side chains exhibit a glass transition temperature; 95 and 80°C for **P5** and **P6**, respectively. For **P1** through **P4**, glass transition has only been observed when the carbazole unit is copolymerized with the quaterthiophene. This might be an important drawback for these polymers, since it will be difficult to form a highly ordered film. The UV-vis spectroscopy has shown very interesting results for these copolymers. As expected, the absorption maxima increase with the number of thiophene moiety in the repeating unit. The only exception being **P2** and the maximum is at lower wavelength because of the steric hindrance between the two thiophene units. The octyl chains prevent the planarity in the molecule and therefore lowering the λ_m and increasing E_g. These measurements of spin-coated thin-films on quartz plates allowed to observe the effect of perfluorinated side chains on the optical properties. As already observed in the literature, these chains decrease the λ_m of about 30 nm for **P5** and **P6**.*[12]* All the copolymers showed very large bandgap ranging from 2.4 to 2.8 eV. Finally, cyclic voltammetry was also performed in the solid state to observe the oxidation potentials of the molecules and to calculate the HOMO energy levels. Since irreversible oxidation by the oxygen in air occurs approximately at -5.2 eV, the polymers should be very stable with their HOMO levels all below -5.4 eV.

Powder XRDs were analyzed at different temperatures to observe the effect on the level of cristallinity during the heating of the polymers. There are two completely different behaviours that were observed during the heating processes of the copolymers. As shown in Figure 3, some of them exhibit some cristallinity by increasing the temperature. For example, **P3** has a much higher level of cristallinity at temperatures between 100 and 150°C. These data should help optimizing the fabrication of field-effect transistors because we then know the most efficient temperatures to be applied on the device after the film deposition. On the other hand, some other copolymers, such as **P1**, **P2** and **P6**, do not exhibit significant changes during the heating process. It is much more difficult to establish the best thermal treatment that has to be applied on these polymers to obtain the best field-effect mobility.

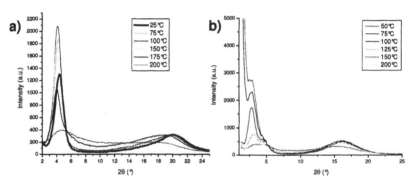

Figure 3. Powder XRD of compounds a) **P3** and b) **P5** at different temperatures.

Device performances

After complete characterization of the copolymers, we were able to fabricate OFETs from spin-coated films on plain silicon dioxide (SiO_2) and octadecylsilane (OTS) modified SiO_2. The devices were completed by evaporating gold electrodes on top of the films. Surprisingly, it was very difficult to cast a homogenous film on the surface even though very high molecular weights were obtained for some polymers. This is the main reason why we could not observe any field-effect for polymers **P2** and **P5**. The source-drain current (I_{DS}) versus source-drain voltage (V_{DS}) at various gate voltages (V_G) and the transfer characteristics for polymer **P1** are shown in Figure 4.

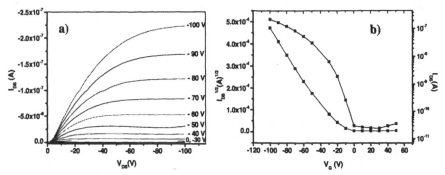

Figure 4. a) Source-drain current (I_{DS}) versus source-drain voltage (V_{DS}) at various gate voltages (V_G) for top-contact field effect transistors using compound **P1** on OTS-treated SiO$_2$. **b)** The transfer characteristics in the saturation regime at a constant source-drain voltage (-100 V) are also shown.

Polymer **P1** gave the best performances, not only did we obtain the highest field-effect mobility but also it is the only polymer which allowed an I_{ON}/I_{OFF} above 10^2 and a negative V_T. The best single device that we have fabricated was with an OTS-modified substrate; a maximum hole mobility of 2 x 10^{-4} cm^2/(V.s), a I_{ON}/I_{OFF} ratio of approximately 10^4 and a V_T of -20 V. This is quite surprising because **P1** is not considered a regioregular copolymer because the bithiophene comonomer is not fully symmetric. The performances were about one order of magnitude higher when we used OTS-treated SiO$_2$ substrate instead of plain SiO$_2$. Unfortunately, the performances were not significantly higher. Even though we showed that the copolymers should have good air stability, the I_{ON}/I_{OFF} was very low. All the performances are described in Table 2.

Table 2. Field-effect mobility, I_{ON}/I_{OFF} current ratio and threshold voltage (V_T) of spin-coated films of **P1** to **P6** with differently treated SiO$_2$ substrates and gold electrodes.

Device	Plain			OTS		
	μ	I_{ON}/I_{OFF}	V_T	μ	I_{ON}/I_{OFF}	V_T
	cm^2/(V.s)		V	cm^2/(V.s)		V
P1	5 x 10^{-6}	2 x 10^2	-3	2 x 10^{-4}	7 x 10^3	-20
P2	No field-effect					
P3	6 x 10^{-6}	60	11	6 x 10^{-5}	4 x 10^2	2
P4	1 x 10^{-5}	20	17	6 x 10^{-5}	40	13
P5	No field-effect					
P6	No field-effect			6 x 10^{-5}	1 x 10^3	-55

We tried several thin-film deposition conditions with different film thicknesses, polymer concentrations and thermal treatments after the deposition. We were unable to obtain better performances with these copolymers. We do not fully understand why the hole mobility is so low, especially for polymers such as **P3** and **P4**. These polymers have high molecular weights and good cristallinity as shown by the XRD. Another intriguing fact about these two polymers is that their threshold voltages are much higher than 0 V. this could be explained by the presence of some impurities in the polymer. Because of that, we believe that these two polymers need to be studied furthermore to improve the thin film organization and increase the hole mobility to reasonable values. The substitution of the regular alkyl chains with perfluorinated chains does not seem to improve the field-effect mobility at all.

CONCLUSION

Finally, we have successfully synthesized several new thiophene-based comonomers. Some of them bearing regular alkyl chains and other perfluorinated alkyl chains to observe the effect of having fluorinated side chains on the polymer. We supposed that we would be able to improve the organization within the polymer and that it would help having better charge transport properties. After the complete characterization, we found that these polymers had very promising properties such as good stability and interesting organization. However, we could not obtain hole mobility higher than 10^{-4} cm^2/(V.s). Optimization of the device fabrication did not lead to any significant improvement. The best performances were obtained with **P1**, a random copolymer, which is very intriguing. It seems that the strategy of adding fluorinated side chains on one of the comonomers did not lead to significant improvement of the performances. Other strategies, such as the addition of fluorine atoms directly on the conjugated backbone, could lead to a better organization.

REFERENCES

[1]   H. Sirringhaus, N. Tessler, R. H. Friend, *Science* **1998**, *280*, 1741.
[2]   H. N. Tsao, D. Cho, J. W. Andreasen, A. Rouhanipour, D. W. Breiby, W. Pisula, K. Müllen, *Adv. Mater.* **2009**, *21*, 209.
[3]   H. Pan, Y. Li, Y. Wu, P. Liu, B. S. Ong, S. Zhu, G. Xu, *J. Am. Chem. Soc.* **2007**, *129*, 4112.
[4]   J. Liu, R. Zhang, G. Sauvé, T. Kowalewski, R. D. McCullough, *J. Am. Chem. Soc.* **2008**, *130*, 13167.
[5]   I. Osaka, R. Zhang, G. Sauvé, D.-M. Smilgies, T. Kowalewski, R. D. McCullough, *J. Am. Chem. Soc.* **2009**, *131*, 2521.
[6]   P.-L. T. Boudreault, S. Beaupré, M. Leclerc, *Polym. Chem.* **2010**, *DOI: 10.1039/b9py00236g*
[7]   S. Cho, J. H. Seo, S. H. Park, D.-Y. Kim, S. Beaupré, M. Leclerc, K. Lee, A. J. Heeger, *Adv. Mater.* **2010**, *DOI:10.1002/adma.200903420*.
[8]   X. Jiang, Y. Harima, K. Yamashita, A. Naka, N.-K. Lee, M. Ishikawa, *J. Mater. Chem.* **2003**, *13*, 785.
[9]   A. Facchetti, M.-H. Yoon, C. L. Stern, G. R. Hutchison, M. A. Ratner, T. J. Marks, *J. Am. Chem. Soc.* **2004**, *126*, 13480.
[10]   A. Facchetti, M. Mushrush, M.-H. Yoon, G. R. Hutchison, M. A. Ratner, T. J. Marks, *J. Am. Chem. Soc.* **2004**, *126*, 13859.
[11]   N. Blouin, A. Michaud, M. Leclerc, *Adv. Mater.* **2007**, *19*, 2295.
[12]   B. Wang, S. Watt, M. Hong, B. Domercq, R. Sun, B. Kippelen, D. M. Collard, *Macromolecules* **2008**, *41*, 5156.

Mater. Res. Soc. Symp. Proc. Vol. 1270 © 2010 Materials Research Society   1270-II09-16

Novel Silylethynyl Substituted Pentacenes with High-Temperature Thermal Transitions

David H. Redinger[1], Robert S. Clough[1], James C. Novack[1], Gregg Caldwell[1], Marcia M. Payne[2], and John E. Anthony[2]

[1]Corporate Research Materials Lab, 3M Company, St. Paul, MN 55144, U.S.A.
[2]Outrider Technologies, LLC, Lexington, KY 40506, U.S.A.

ABSTRACT

Modifications to the p-type semiconductor TIPS-Pentacene can result in elimination of the solid-solid thermal transition at 124 °C. This new material has shown mobility higher than 1 cm^2/Vs. Elimination of the solid-solid thermal transition leaves the melting point as the lowest temperature transition at 199 °C.

INTRODUCTION

Over the past several years 6,13-bis(triisopropylsilylethynyl) pentacene (TIPS-Pentacene) has been extensively studied for use in organic thin-film transistors [1], with display backplanes as a leading application. The soluble nature of TIPS-Pentacene enables solution-processing techniques which may result in a lower cost of fabrication [2]. Several groups have demonstrated thin-film transistors (TFTs) based on TIPS-Pentacene with mobility in excess of 1 cm^2/Vs [3,4]. Although device performance for this material is generally quite good, TIPS-Pentacene has a well-known solid-solid thermal transition at approximately 124 °C [5], which leads to cracking in crystalline films along with a decrease in measured hole mobility. This transition has the potential to limit its use in applications that require high-temperature processing such as photoresist baking or color filter lamination. Elimination of the solid-solid thermal transition would allow small molecule organic semiconductors to be compatible with a wide range of higher temperature processes, and ultimately make the glass-transition temperature of a flexible polymeric substrate the limiting factor in device fabrication.

Small modifications to the TIPS-Pentacene molecule can result in significant changes in semiconductor properties. Substitution of only one of the three isopropyl groups with another group can lead to large changes in solubility, mobility, crystal packing (e.g. 1-D versus 2-D π-facial intermolecular interactions (π-stacking)) and thermal transition characteristics. The goal of this work was to identify novel semiconducting materials, based on modifications of the silyl-substituted groups of TIPS-Pentacene, which provide benefits in processing and higher performance.

EXPERIMENT

Over 50 new materials have been investigated with only a few identified as promising candidates for thin-film transistors. Given the large number of materials a systematic method of evaluation was necessary. Crystallographic information was obtained using XRD on single crystals which determined whether the material was a 1D or 2D π-stack motif. Generally, materials with 2D π-stacking are better suited for TFTs, and therefore these materials were subject to further evaluation. Thermal transitions were identified using differential scanning calorimetry (DSC). If a sufficient amount of material was available solubility was determined using toluene as a standard solvent. Although solubility of each material differs from TIPS-Pentacene, all materials mentioned here are soluble to at least 1-2 wt.% in common coating solvents such as n-butylbenzene, anisole, and toluene. Finally, TFTs were fabricated to determine the mobility of the semiconductor. In this work a conventional bottom-gate top-contact architecture was chosen for its simplicity. While not an optimal design, it gave consistent results and allowed differentiation between semiconductor materials without interference from process variation. Dip-coating was chosen as the semiconductor deposition method because of the relatively good performance and excellent control. The substrate was a heavily doped n-type silicon wafer (As) with 1000 Å of thermal oxide on the front surface and coated with 100 Å TiN and 5000 Å aluminum on the back surface. Prior to coating the wafer was oxygen plasma cleaned for three minutes. Each substrate sample was coated using draw rate of 3-5 millimeters per minute. Approximately 5 mL of each solution was used to fill the dip-coating tank, which required approximately 100 mg of semiconductor for a typical 2 wt.% solution. Each semiconductor solution was filtered through a 0.2 micron PTFE filter prior to use. Samples were allowed to dry at room temperature under atmospheric conditions. After coating long crystals were present on the SiO_2 surface, typically oriented parallel to the dip axis. Gold source and drain electrodes (approximately 800-1000 Å thick) were deposited via thermal evaporation through a shadow mask. Source and drain contacts were oriented relative to the semiconductor crystals such that the crystals bridged the contacts. All channel lengths were 100 μm.

DISCUSSION

Solid-solid thermal transitions are thought to limit the maximum processing temperature, and long-term exposure of TIPS-Pentacene devices to temperatures above 150 °C can result in reduced TFT performance [6]. However, little is known regarding the impact of this transition on long-term stability. In large crystal form, high temperature excursions can cause cracking of the material. However, thin-films of TIPS-Pentacene have been subjected to temperatures of 145 °C (21 °C above the solid-solid transition at 124 °C) for 30 minutes without suffering visible damage or reduction in TFT mobility. While these results indicate process temperatures above the solid-solid transition temperature are not immediately fatal to TFTs, elimination of this transition would be desirable.

Figure 1 shows DSC thermograms for TIPS-Pentacene and L-21081, a material which has a lower solid-solid thermal transition temperature (shown by the endotherm at approximately 70 °C). The structure of L-21081 allows the position of each silyl substituent to be uniquely identified in the crystal structure. XRD results taken at 70 °C indicate a rotation of the silyl groups and increased disorder within the crystal. There is also a slight increase in inter-molecular spacing. This solid-solid transition is reversible and the heat can be recovered upon the reverse scan. The exotherm near 200 °C is not reversible, and is likely a result of a Diels-Alder based degradation.

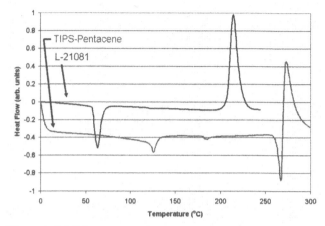

Figure 1. DSC thermogram of TIPS-Pentacene and L-21081. L-21081 shows a solid-solid transition at approximately 70 °C.

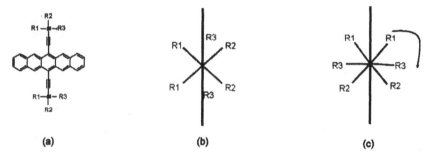

(a) (b) (c)

Figure 2. At low temperatures the positions of the silyl substituted groups are fixed (b). At the solid-solid thermal transition temperature the groups are allowed to rotate, resulting in increased disorder within the system.

RESULTS

Solid-solid thermal transitions can be eliminated altogether by choosing the right set of silyl substituents. Figure 3 shows a DSC thermogram of L-21080, which shows no indication of a solid-solid transition. The exotherm at approximately 199 °C is irreversible and corresponds to the degradation of the material. Thus this material may offer an advantage over TIPS-Pentacene in applications which require processing temperatures in excess of 124 °C, provided proper precautions are taken with respect to thermal coefficients of expansion and stress.

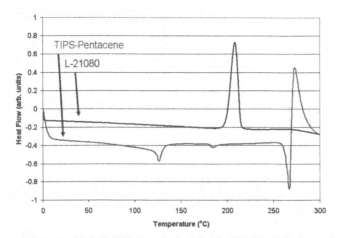

Figure 3. DSC thermogram of L-21080 compared with TIPS-Pentacene.

The measured mobility of L-21080, even in the non-optimized bottom gate TFT process, is approximately 1 cm^2/Vs. For comparison, TIPS-Pentacene typically shows mobility in the 0.3-0.4 cm^2/Vs range using a similar process. Figure 4 shows dip-coated crystals of L-21080 on the SiO$_2$ substrate. An Id-Vg curve taken in the saturated regime is shown in Figure 5.

Figure 4. Dip-coated crystals of L-21080 showing a high degree of order on the surface of the substrate.

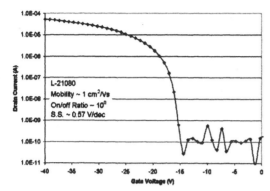

Figure 5. Id-Vg of a TFT using L-21080 with mobility of approximately 1 cm²/Vs.

CONCLUSIONS

A new soluble small molecule organic semiconductor based on modifications to TIPS-Pentacene has been developed which does not undergo a solid-solid phase transition. The observed TFT mobility was higher than TIPS-Pentacene in a comparable device construction, and may be further improved through process optimization. These results show that solid-solid transitions can be eliminated without sacrificing mobility, solubility, or processability.

REFERENCES

1. J. Anthony, J. Brooks, D. Eaton, and S. Parkin, *J. Am. Chem. Soc.* **123**, 9482–9483 (2001).
2. H. Sirringhaus, T. Kawase, R.H. Friend, T. Shimoda, M. Inbasekaran, W. Wu, and E.P. Woo, *Science* **290**, 2123-2126 (2000).
3. S.K. Park, T.N. Jackson, J.E. Anthony, and D.A. Mourey, *Applied Physics Letters* **91**, 063514 (2007).
4. J. Halls, C. Newsome, T. Kugler, G. Whiting, C. Murphy, J. Burroughes, "OTFT development for OLED backplanes: optimisation of high mobility 10 μm channel OTFTs," Presented at Intertech Thin-Film Transistors Conference (2008).
5. J. Chen, J. Anthony, and D. Martin, *J. Phys. Chem. B* **110**, 16397–16403 (2006).
6. J. Chen, C. K. Tee, J. Yang, C. Shaw, M. Shtein, J. Anthony and D. Martin, *J. Polym. Sci. B: Polym. Phys.* **44**, 3631 – 3641 (2006).

Mater. Res. Soc. Symp. Proc. Vol. 1270 © 2010 Materials Research Society   1270-II09-19

Sensitizer Effects on the Transport Properties of Polymer:Sensitizer Organic Blend

Karina Aleman[1], Svetlana Mansurova[1], Andrey Kosarev[1], Ponciano Rodriguez[1], Klaus Meerholz[2] and Sebastian Koeber[2]

[1]National Institute for Astrophysics, Optics and Electronics, AP 51 y 216, Puebla, 72000, Mexico
[2]University of Cologne, Insitute of Physical Chemistry, Luxemburger Str. 116, Cologne, D-50939, Germany

ABSTRACT

In this work we studied the effect of sensitizer concentration on a mobility-lifetime product $\mu\tau$, on photoconductivity response time τ_{ph}, and on drift mobility μ of the majority carriers in an organic polymer:sensitizer blend. The intensity modulated photocurrent and photo-EMF technique were used as experimental tools for this purpose. The studied material consists of a mixture of the novel non-conjugated main chain hole-transporting polymer PFO6:PDA (Poly(N,N'-bis(4-hexyloxyphenyl)-N'-(4-(9-phenyl-9H-fluoren-9-yl)phenyl)phenylen-1,4- diamine) sensitized with the highly soluble C_{60} derivative PCBM (phenyl-C_{61}-butyric acid methyl ester) in the range from $Z = 1$ to 40 wt.-%. It was experimentally observed that (1) at the increasing sensitizer concentration the overall photoconductivity increases; (2) the majority carrier type switches from holes to electrons at approximately 2:1 polymer:sensitizer ratio; (3) the holes response time becomes shorter at the decreasing polymer fraction, while the electrons lifetime is only slightly dependent on sensitizer concentration; (4) the hole mobility-lifetime product decreases at the decreasing concentration of hole transporting component (polymer), while the electrons mobility-lifetime product increases at the increasing concentration of the electron transporting component (sensitizer); (5) the same is true for the carriers mobilities.

INTRODUCTION

Due to their promising properties such as flexibility, possibility to use printing process in roll to roll and low cost, organic solar cells have attracted a lot of attention during the last decade. Bulk heterojunction (BHJ) solar cells based on blends of donor (photoconductive polymer) and acceptor (small molecule sensitizer) belong to the most promising candidates for this application, as efficiencies in the range of 5% have been achieved recently [1] in poly(3-hexyl thiophene-2,5-diyl) (P3HT):(6,6)-phenyl-butyric acid methyl ester (PCBM) blend devices. Transport properties of the active material are the crucial factors determining the performance of organic solar cells [2, 3]. However, as the blend of two different material types cannot be described by a simple superposition of the single material properties, a prediction of the bipolar charge transport in a solar cell is not straightforward. In this context, the experimental data on donor/acceptor ratio dependent transport characteristics would have been of particular interest and a considerable amount of work has been devoted recently to this problem [4-7]. The experimental methods used to obtain information about charge transport in organic semiconductors – time-of-flight (TOF), transient photoconductivity, charge extraction by linearly increasing voltage (CELIV), current-voltage measurements in space charge limited current regime, and field effect transistor (FET) measurements are mostly focused on determination of charge carrier mobility. There is only one publication studying the influence of the varying donor:acceptor ratio on the carrier's mobility-lifetime product in the organic blend. Dennler et al. [8] measured the majority charge carrier mobility and bimolecular lifetime in poly[2- methoxy-5-(3',7'- dimethyloctyloxy)-1,4-phenylenevinylene (MDMO-PPV) and PCBM mixtures as a function of the concentration using CELIV technique at room temperature. It was found that the carrier's mobility is increasing with increasing PCBM concentration, whereas the carrier lifetime is decreasing, producing ratio independent mobility-lifetime product.

Here we study the influence of sensitizer concentration on the majority carriers transport properties in novel material consisting of a mixture of non-conjugated main-chain hole-transporting polymer PFO6:PDA [9] sensitized with the highly soluble C_{60} derivative PCBM. The characterization was performed using a method known in current literature as non-steady-state photo-EMF (p-EMF) technique [10], which consists on measurement of the alternaning current (AC) excited by a spatial and temporal modulation of the excess carriers induced by the vibrating interference pattern in short circuited photoconductive sample. It was shown recently [11] that non-steady-state photo-EMF tecnique can be succesfully applied for characterization of organic (polymer) semiconductors) allowing to measure a number of important material parameters such as mobility-lifetime $\mu\tau$ product, photo carriers lifetime τ, dielectric relaxation time τ_{di}, primary quantum efficiency of charge generation ϕ, etc.. The additional motivation for chosing photo-EMF technique as charactrization method of polymer:sensitizer blend comes from the fact that characterization can be performed at the experimental condition (electric field, light intensities, etc.) close to the device (solar cells) operational one.

EXPERIMENTAL DETAILS

Experimental method: A detailed description of the non-steady state photo-EMF technique can be found, e.g., in [10]. Illumination of the photoconductive sample with a sinusoidal interference pattern gives rise to a grating of photoexcitied mobile carriers. Redistribution of the mobile charges through drift or diffusion and their trapping on localized states results in a spatially inhomogeneous space-charge field E_{sc} inside the sample. If one of the beams is modulated in phase, the interference pattern and as a consequence the mobile carrier grating is oscillating. The interaction of the relatively stable space-charge-field grating with the moving free-carrier grating generates an alternating current (AC) that flows through the short-circuited sample. If the external DC bias is applied to the sample the current density amplitude is given by a following expression:

$$j_{p-emf}^{\Omega} \propto \sigma_0 \frac{E_D(1+K^2L_D^2)-KL_0E_0}{(1+K^2L_D^2)^2+K^2L_0^2} \qquad (1)$$

Here σ_0 is an average photoconductivity, $E_D = K(D/\mu)$ is so called diffusion field, $K = 2\pi/\Lambda$ is spatial frequency of the interference pattern and Λ is a period of interference fringes. Measuring the dependences of the photo-EMF current J_{p-emf}^{Ω} on the external DC field E_0, the drift lengths of the photo carriers can be evaluated. From Eq. (1), it is seen that the photo-EMF signal amplitude drops to zero when the diffusion and drift components of the space charge grating compensate each other, i.e. when the condition: $E_D(1+K^2L_D^2) = KL_0E_0'$ is met. Assuming the validity of the Einstein relation between the mobility and diffusion coefficient $D/\mu = k_B T/e$, the mobility-lifetime product $\mu\tau$ can be evaluated as $\mu\tau = (k_B T/e)(E_0^{'})^2$ from the position of the minimum $E_0^{'}$.

Samples: Samples with a thickness of 8 μm and 0.25 cm² front/back area were prepared by sandwiching the material between two glasses covered by the transparent indium tin oxide (ITO) electrodes. The studied material consists of a mixture of the novel non-conjugated main-chain hole-transporting polymers polymer PFO6:PDA sensitized with the highly soluble C_{60} derivative PCBM and chronophers 2,5-dimethyl-(4-p-nitrophenylazo)-anisol (DMNPAA) and 3-methoxy-(4-p-nitrophenylazo)-anisol (MNPAA). The chronopher concentration was fixed as 50 wt.- % in all samples. Six samples with the varying sensitizer concentration Z = 1, 5, 10, 15, 25, 40 wt.- % where prepared.

Experimental arrangement: The photo-EMF experiments were performed in two different configurations. For the first four samples (1-15 wt %) the reflection configuration with two counter-propagating beams illuminating the sample from the opposite sites was chosen. The period of the interference pattern for this configuration was $\Lambda \approx 0.2$ μm (for the refraction index value $n = 1.7$). The last two samples (25 and 40 wt. %) were measured in transmission geometry in this case two beams illuminating the sample on the same side. The beam crossing angle in transmission configuration was 40° and tilt angle was 45°, creating interference pattern with the spatial period $\Lambda \approx 0.93$ μm. The 633 nm output of a cw He-Ne laser was used to illuminate the samples with an interference pattern of high fringe contrast ($m \approx 1$). The phase of one of the beams was periodically modulated by an electro-optic modulator (Conoptics 350-105) with amplitude of modulation $\Delta = 0.5$ rad. The total incident intensity was 0.8 mW/cm². Conventional intensity modulated photocurrent measurements were performed by blocking one of the two beams and by modulating the intensity of other beam by an electro optical modulator. The output signal was detected by a digital lock-in amplifier (Stanford Research SR-830).

RESULTS AND DISCUSSION

The AC photocurrent amplitude as a function of the modulation frequency is shown in figure 1. The signal is frequency independent at low modulation frequencies and then starts to decay after some characteristic frequency Ω_{ph}. The important observation is that Ω_{ph} increases at higher sensitizer content (figure 2), indicating that the majority carrier response time $\tau_{ph} = \Omega_{ph}^{-1}$ becomes shorter. The photoconductivity response becomes faster by one order of magnitude (from $\tau_{ph} = 10.6$ ms at Z = 1 wt.- % to $\tau_{ph} = 1$ ms at Z = 40 wt.- %). The electric field dependences of the AC photocurrent is depicted in figure 3. The signal amplitude increases as the sensitizer concentration grows, indicating the increase of the photoconductivity. The photocurrent depends on electric field as $I_{ph}(E) \approx AE^\gamma$. There are two regions in the curves: the first region at low field (ranging from $E_0 = 1.25$ $V\mu m^{-1}$ to $E_0 = 5$ $V\mu m^{-1}$) shows a super-linear behaviour characterized by γ_1. At larger DC field, the second region with greater γ_2 occurs. In figure 4, the super-linear parameters for both regions are shown.

The photo-EMF signal J_{p-emf}^{Ω} dependence of on the frequency of modulation Ω

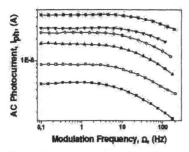

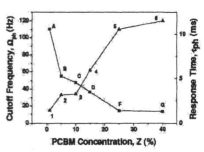

Figure 1. AC photocurrent amplitude vs. modulation frequency for different sensitizer concentration as $Z = 1\%$ (■), 5% (□), 10% (▲), 15% (○), 25% (▼), 40% (*).

Figure 2. The characteristic frequency Ω_{ph} (▲) and response time τ_{ph} (■) vs. the PCBM concentration.

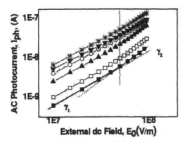

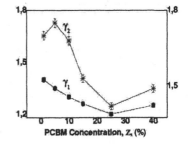

Figure 3. AC Photocurrent as a function of electric field for different sensitizer concentration $Z = 1\%$ (■), 5% (□), 10% (▲), 15% (○), 25% (▼), 40% (*).

Figure 4. The super-linear behaviour at low electric field γ_1 (■) and high field γ_2 (*) as a function of Z.

is shown in figure 5 for different sensitizer concentration. J^{Ω}_{p-emf} grows as the modulation frequency increases until a characteristic cut-off frequency $\Omega_c = 1/\tau_{p-EMF}$ determined by the photo-EMF response time τ_{p-EMF}. The values of the Ω_c and τ_{p-EMF} as a function of Z are presented in figure 6. It can be observed than τ_{p-EMF} decreases as the increasing sensitizer fraction. The important observation is that in high frequency region ($\Omega > \Omega_c$) there was an abrupt change on the phase of the photo-EMF signal by approximately 180° between two group of the samples with low (Z1 = 1 - 15 wt.-%) and high (Z2 = 25 and 40 wt.-%) sensitizer concentrations, indicating the change of the majority carrier type [10]. Comparing the signal sign with that of the reference sample, the majority carriers for Z1 group were identified as holes, while for Z2 they were electrons. This is not surprising, if one takes into account the well known in a current literature fact, that polymer component is responsible for the conduction of holes and sensitizer for the conduction of electrons. The transition from electron to hole conductivity in our material occurs at approximately 2:1 polymer:sensitizer ratio.

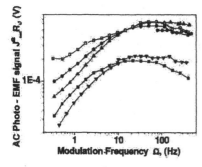

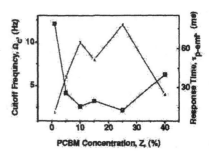

Figure 5. Photo- EMF signal vs. modulation frequency for different Z = 1% (▼), 5% (■), 10% (▲), 15% (●), 25%(□), 40% (*).

Figure 6. Cutoff frequency Ω_c (□) and photo-EMF response time τ_{p-EMF} (▲) vs. PCBM concentration.

Figure 7 represents typical dependences of the photo-EMF current on the external DC field at different sensitizer content. It is seen that the photo-EMF signal has a well pronounced minimum at low DC field, and then increases with growing E_0. From the minimum position the product μt was determined. Using the data on the mobility-lifetime product $\mu\tau$ and conductivity response time τ_{ph}, the drift mobility μ_{dr} was calculated. The dependence of both transport parameters on Z is shown in figure 8.

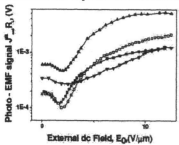

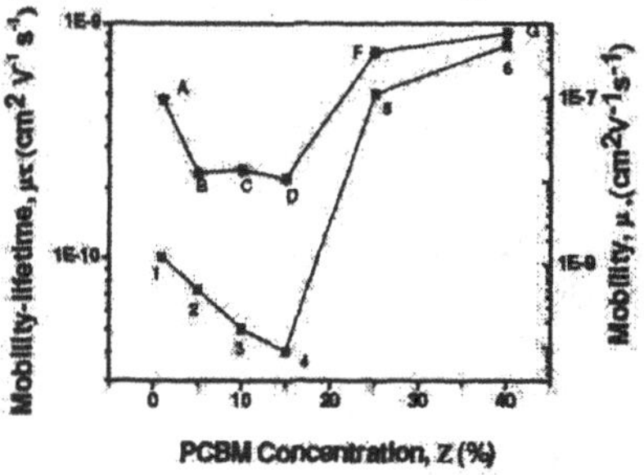

Figure 7. Photo-EMF signal vs. external DC field for different Z = 1% (▼), 5% (■), 10% (▲), 15%(●).

Figure 8. Mobility-lifetime $\mu\tau$ (■) and drift carrier mobility μ (*) as a function of the PCBM sensitizer concentration Z.

It can be observed (figure 8) that in Z1 group the holes transport length decreases at the decreasing polymer fraction, and in Z2 group the electrons transport length increases at the increasing sensitizer fraction. For the Z2 group the $\mu\tau$ and μ_{dr} increases drastically compared with Z1 group. Next, we will briefly discuss our main experimental findings. The strong dependence of electron and hole motilities on the concentration of corresponding transporting component (figure 8) is in agreement with the experimental data on similar polymer:sensitizer mixtures (see e.g. classical PVK:TNF study by Gill [11]). This behavior can be explained in a framework of theory of hopping charge transport via localized states [12], which predicts an exponential dependence of the mobility on the distance between two jumping sites. As to the shorter photoconductivity holes response time τ_{ph} (figure 2) at the decreasing concentration of polymer fraction, this can be an indication of an increasing concentration of recombination centers due to the growing number of "dead-ends" when the connection between the neighboring polymer domains is lost. Finally we will comment on the observed growth of photoconductivity with the increasing sensitizer content. In our opinion, the reasonable explanation for this behavior is the possible interplay between different trends on quantum efficiency of charge generation and the majority carriers transport length. Indeed, at low-to-medium Z (Z1group) the growing quantum efficiency of charge generation due to the increased effective area of the donor-acceptor interface (and as a consequence, a more efficient separation of photo excited electron-hole pairs) may be responsible for the observed photoconductivity increase, while at medium-to-high Z (Z2 group) the large electron transport length can start to play a decisive role.

CONCLUSION

The transport properties of novel polymer:senitizer blend were analyzed as a function of components fraction using intensity modulated photocurrent and non-steady state photo EMF technique. The experimental results allows to conclude: (1) at the increasing sensitizer concentration the overall photoconductivity increases; (2) the majority carrier type switches from holes to electrons at approximately 2:1 polymer:sensitizer ration; (3) the holes response time becomes shorter at the decreasing polymer fraction, while the electrons lifetime is only slightly dependent on sensitizer concentration; (4) the hole mobility-lifetime product decreases at the decreasing concentration of hole transporting component (polymer), while the electrons mobility-lifetime product increases at the increasing concentration of the electron transporting component (sensitizer); (5) the same is true for the carriers mobilities.

ACKNOWLEDGMENTS

Authors acknowledge CONACyT project No.84922 for financial support, Karina Aleman acknowledges CONACyT scholarship No. 207900.

REFERENCES

1. R. Green, A. Marfa, A. J. Ferguson, N. Kopidakis, G. Rumbles, and S. E. Shaheen, Appl. Phys. Lett. **92**, 033301 (2008).
2. C. Deibel, A. Wagenpfahl, and V. Dyakonov, Phys. Stat. Sol. (RRL) **2**, 175 (2008). ,
3. A. Pivrikas, N. S. Sariciftci, G. Jûska, and R. Österbacka Prog. Photovolt: Res. Appl., **15** (2007)].
4. S. M. Tuladhar, D. Poplavskyy, S. A. Choulis, J. R. Durrant, D. D. C. Bradley, and J. Nelson, Adv. Funct. Mater. **15** , 1171 (2005).
5. E. von Hauff, J. Parisi, and V. Dyakonov, J. Appl. Phys. **100** 043702 (2006)
6. J. Huang, G. Li, and Y. Yang, Appl. Phys. Lett. **87** , 112105 (2005).]
7. A. Baumann, J. Lorrmann, C. Deibel, V. Dyakonov, Applied Physics Letters, **93**, 252104, 2008]
8. G. Dennler, A. J. Mozer, G. Jûska, A. Pivrikas, R. Österbacka, D. A. Fuchsbauer, and N. S. Sariciftci, Organic Electronics, **7**, 229 (2006)
9. J. Schelter, G. F. Mielke, A. Köhnen, J. Wies, S. Köber, O. Nuyken, K. Meerholz
10. S. Stepanov, *Photo-Electromotive-Force Effect in Semiconductors* in Handbook on Advanced Electronics and Photonics Materials, edited by H. S. Nalwa, (Academic, San Diego 2001).
11. W. D. Gill, J. Appl. Phys. **43**, 5033 (1972).
12. S. D. Baranovski and O. Rubel, Charge transport in amorphous semiconductors. In: *Charge Transport in Amorphous Solids with Application in Electronics*, S. D. Baranovski (ed.) John Willey & Sons, Ltd., Chichester, (2006)

Mater. Res. Soc. Symp. Proc. Vol. 1270 © 2010 Materials Research Society   1270-II09-49

Built-in potential of a pentacene p-i-n homojunction studied by ultraviolet photoemission spectroscopy

Selina Olthof, Hans Kleemann, Björn Lüssem, and Karl Leo
Institut für Angewandte Photophysik, Technische Universität Dresden,
George Bähr Strasse 1, 01062 Dresden, Germany

ABSTRACT

In this paper we investigate the energetic alignment in an organic p-i-n homojunction using ultraviolet photoelectron spectroscopy. The device is made of pentacene and we emploay the small molecules NDN1 for n-doping and NDP2 for p-doping the layers. The full p-i-n structure is deposited stepwise on a silver substrate to learn about the interface dipoles and band bending effects present in the device. From the change in work function between the p- and n-doped layers we gain knowledge of the built-in potential of this junction.

INTRODUCTION

In the last years, organic semiconductors have been investigated intensively due to the interest in their basic physical principles and their large application potential in organic electronics and optoelectronics, allowing devices such as organic light emitting diodes [1] or organic solar cells [2]. Despite the already successful commercial application of such devices, the general state of the art of the field of organic semiconductors is still rather immature. For instance, many basic devices known from inorganic semiconductors have not been realized from organic semiconductors so far.

Until recently, this applied also to the most fundamental semiconductor device of all, the pn-homojunction. A few years ago, Harada et al. [3] reported the first stable and reproducible organic homojunction based on p- and n-doped layers of the organic semiconductor zinc phthalocyanine. The challenge in realizing stable molecularly doped layers for a homojunction lies in the selection of dopant molecules which can dope the same host material both p- and n-type. This requires an n-dopant with very high lying LUMO (lowest unoccupied molecular orbital) and a p-dopant with very high-lying HOMO (highest occupied molecular orbital) [4]. Furthermore, the devices have to be realized in the form of p-i-n-junctions [3]. The intrinsic interlayer is needed to avoid tunnelling due to the rather narrow space-charge layers in doped organic semiconductors [5]. More recently, a p-i-n-homojunction could also be realized in pentacene [6], showing an exceptionally large built-in potential. The experimental results reported in Refs. [5,6] have been explained by the Shockley theory for pn-junctions where modifications due to the hopping transport in organic materials were taken into account.

In this work, we investigate a p-i-n homojunction by ultraviolet photoelectron spectroscopy (UPS) to obtain further information on the basics alignment in the device, including interface dipoles, band bending effects and the built-in potential. As matrix material, we choose pentacene (PEN) since it has excellent hole and electron transporting properties [8] and is a popular material for organic field-effect transistors. As mentioned before, when building a homojunction the challenge is to find dopants that can achieve n- and p-type doping for the same host material: For p-doping, the dopant electron affinity (EA) has to be close to or lager than matrix ionization potential (IP), for n- doping the dopant IE has to be in the range of the matrix EA. To meet both requirements, strong donor and acceptor molecules are needed. In our case, for n-doping NDN1 and for p-doping NDP2 are chosen, which both are proprietary materials provided by Novaled AG.

EXPERIMENT

The samples are prepared under ultra high vacuum conditions. The pentacene (Sensient, 2x sublimated) and the dopants (Novaled AG) are thermally evaporated from crucibles at a pressure of $\leq 5 \cdot 10^{-8}$ mbar and incrementally deposited at room temperature at rates of 6 Å/min for the host and < 0.5 Å/min for the dopants. A doping of 2 mol% is used, which is achieved by co-evaporation of the materials and can be controlled by two separate quartz crystals. In order to avoid cross contamination, the intrinsic and either of the doped layers are evaporated in three different chambers. The setup for UPS measurements is directly connected to the evaporation tool, therefore samples can be transferred without breaking the vacuum. Measurements are performed at a base pressure of $< 5 \cdot 10^{-9}$ mbar with a Phoibos100 system (Specs) using the HeI line at hv=21.22 eV from a helium discharge lamp.

The samples are prepared on a sputter cleaned silver foil (MaTecK 99.995%) and the investigated thicknesses of the different layers range from 2 Å up to 30 nm. For obtaining the true energetic alignment in the actual device, it is not sufficient to prepare the device in only one deposition direction, as it is unknown how the underlying layers react upon deposition of the top layers. Therefore, two samples are investigated. On one sample (sample A) the interfaces silver to p-PEN, p-PEN to i-PEN and finally i-PEN to n-PEN are measured interface resolved. The same device is then built once more in reversed order (sample B) to gain the full information on the alignment.

DISCUSSION

In sample A, p-PEN is deposited on silver followed by intrinsic PEN and finally n-doped PEN. The resulting UPS spectra are shown in Fig. 1 and are vertically shifted for clarity. The binding energy scale is referenced to the Fermi energy of the silver substrate. The left side of the graph shows the normalized high binding energy cutoff (HBEC) that is associated with the work function (Wf) via the equation:

$$Wf = hv - HBEC.$$

On the right side of Fig. 1, the valence band region is displayed with the HOMO cutoff marked by vertical lines and the respective layer thickness of the different layers denoted next to it. The changes in work function and HOMO position as a function of the layer thickness are shown in Fig. 2 a). The freshly sputtered silver foil (lowest curve in Fig. 1) shows a Wf of 4.3 eV. Upon the deposition of the first monolayer of p-PEN (~1 nm), an abrupt shift to higher binding energy of 0.4 eV in the HBEC is observed as an interface dipole is created. The interface dipole can have several reasons, including the metal pillow effect [9] or charge transfer between metal and molecule [10,11]. This shift is reversed for increasing thickness by a gradual level bending towards lower binding energy, clearly indicating a p-doping effect [5,12]. The overall shift of the band bending is 1 eV and takes about 20 nm. This results in a depletion layer thickness of the doped organic layer in this range.

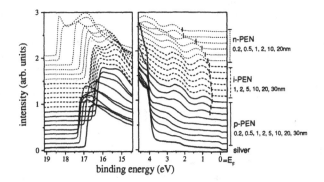

Figure 1: UPS spectra measured on sample A. On top of a silver sample (bottom curve) p-doped PEN is evaporated (solid curves) followed by intrinsic PEN (dashed curves), and n-doped PEN (dotted curves). The layer thicknesses are denoted next to the graph; the left side shows the HBEC and the right side the valence band region with the HOMO cutoffs marked by vertical lines.

The HOMO position of PEN can only be distinguished after 10 nm of coverage and the Fermi edge of the underlying silver can be seen up to this point. This suggests an island-like growth in the beginning as it is commonly observed with PEN [13,14]. The IP of the 30 nm thick p-doped PEN is found to be 5.26 eV and the hole injection barrier (HIB) is 0.38 eV.

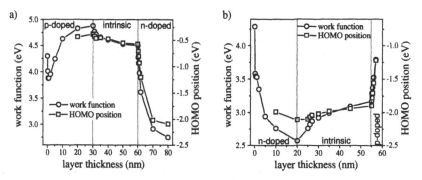

Figure 2: Shift of the HOMO position (squares) and work function (circles) for a) sample A and b) sample B. Vertical lines mark the interfaces between the organic layers.

Upon deposition of the intrinsic PEN, no interface dipole is observed. However, there is a gradual downward shifting of the vacuum level across the 30 nm thick layer by 400 meV, accompanied by a somewhat smaller shift in the HOMO cutoff position of 200 meV. The dissimilar shift of these two values is due to a change in the ionisation potential, as the intrinsic layer shows a lower value of IP = 5.07 eV. After 30 nm of intrinsic PEN, the Wf has a value of

197

4.5 eV and HIB = 0.57 eV, but no saturation of the shift is yet achieved. The measurement is not carried on, as this is a typical thickness of the intrinsic layer used in a PEN homojunction device.

On top of this intrinsic layer, n-doped PEN is evaporated. Immediately after 0.2 nm coverage, a strong downward shift is observable in the vacuum level as well as in the HOMO position. Since the HOMO signal at that point still originates from the underlying intrinsic PEN, it is obvious that not only the n-PEN is shifting, but rather the intrinsic layer is pulled downward at the interface to the doped layer. The shifting saturates after about 10 nm coverage with a total change in vacuum level energy of 1.7 eV and in HOMO position of 1.52 eV. At 30 nm thickness, the n-doped layer reaches IP = 4.84 eV and HIB = 2.09 eV.

The deposition of the n-doped layer does not only have an influence on the alignment of the intrinsic layer, but on the p-doped layer as well. This can be observed by a change in position of a characteristic core level peak of the p-dopant by XPS that is still visible during the deposition of the n-PEN. The p-layer layer is pulled downward at the interface to the intrinsic layer when the depletion layer forms. This change in position accounts for 780 meV. As NDP2 is a proprietary material, the data can not be shown.

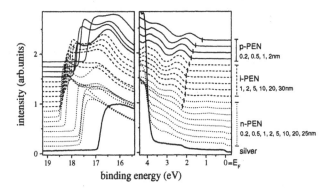

Figure 3: UPS spectra of sample B. On top of a silver sample (bottom curve) n-doped PEN is evaporated (dotted curves) followed by intrinsic PEN (dashed curves), and p-doped PEN (solid curves). The layer thicknesses are denoted next to the graph; the left side shows the HBEC and the right side the valence band region with the HOMO cutoffs marked by vertical lines.

For the second sample B, the deposition sequence is reversed, starting with the n-doped PEN layer on silver, followed by the intrinsic and p-doped layers. The resulting spectra are plotted in Fig. 3 and the shifts of the work function and HOMO position are displayed in Fig. 2 b). The freshly sputtered silver foil again shows a Wf of 4.3 eV. With deposition of the first sub-monolayer of n-PEN on Ag, a strong downward shift in the vacuum level is observed as again an interface dipole is created. This shift is now followed by a downward level bending, i.e., this time both shifts go in the same direction and are therefore not distinguishable. Only the total change can be stated to be 1.53 eV. After about 10 nm, the level bending saturates which is faster than for the p-side layer, indicating a more efficient doping by NDN1. The HOMO cutoff can be seen after 10 nm coverage and again the Fermi edge is visible until then, as was the case for the p-doped layer, so the growth modes of p and n doped layers seem to be the same. The ionization potential of the 25 nm thick layer is 4.85 eV and HIB = 2.1 eV as it was found for the topmost layer in sample A. This reproducibility proves that in sample A, the equilibrium alignment has indeed been reached for the top layer and that the doped layers have a fixed Fermi energy position in their band gap, independent on the underling substrate [18]. This is not the case for intrinsic semiconductors [19,20].

When the intrinsic layer is evaporated on top, the vacuum level aligns with the n-doped layer and therefore starts off at a different energetic position as on sample A. Throughout the intrinsic layer, a gradual upward shift of the vacuum level by 410 meV is observed, accompanied by a smaller 200 meV shift in the HOMO as again the IP is slightly changing. After 30 nm of coverage, again an IP of 5.1 eV is reached, now at HIB = 1.9 eV.

Upon deposition of the p-type layer, a strong upward shift is seen in the vacuum level and HOMO position. Again this shift happens mainly in the intrinsic layer, even though a characteristic core level peak of NDN1 indicates an upward shifting by 270 meV of the n-doped layer as well. The measurement could only be done upon 2 nm p-PEN coverage, since for thicker layers the sample starts to get charged. This is expected since in this direction the sample acts as a blocking device for the holes created in the photoemission process. At this point, the vacuum level shift by 0.6 eV but the bending is not completed yet. Just as in the previous case, it can be assumed that the final position of the p-doped layer will be the same as measured directly on the silver substrate in sample A.

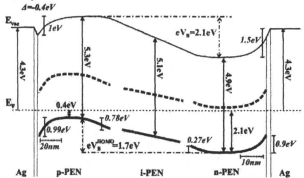

Figure 4: Schematic device alignment of the p-i-n pentacene homojunction concluded from the UPS measurements of samples A and B. Two built-in voltages are stated, one is calculated by the difference in work function (V_B) and the other one by the difference in hole injection barrier (V^B_{HOMO}). The value of the transport gap of 2.7eV in the graph is taken from literature [14].

CONCLUSIONS

Figure 4 summarizes the results of the forward and backward measurement of the p-i-n homojunction and displays the alignment of the full device. Here, we assume the ideal case that a silver top contact on n-PEN would create the same energetic alignment as the measured bottom contact. This is most likely not the case [21] but in this work we are mainly interested in the organic interfaces and their built-in potential and therefore omit this point.

The p- and n-layers show the characteristics of doping, namely forming a depletion region in contact to the metal and a Fermi energy close to the transport level. The built-in potential V_B of a device is usually given by the difference in work function of the doped layers, which accounts for $V_B = 2.1$ eV. However, of more interest for the device performance is the voltage needed to achieve a flatband condition. Because the ionization potentials of the p-doped, n-doped, and intrinsic layers differ, the built-in voltage of the transport levels is only $V^B_{HOMO} = 1.7$ eV. Thereof, 0.65 eV drop across the intrinsic layer and 1.05 eV drop across the depletion layers of the doped pentacene.

ACKNOWLEDGMENTS

This work was funded by the German BMBF under Contract No. 13N 8855, project acronym "ROLLEX". Furthermore, the authors want to thank the Noavaled AG for providing the dopants NDP2 and NDN1

REFERENCES

[1] C. W. Tang and S. A. VanSlyke. Appl. Phys. Lett. 51 (1987) 913.
[2] G. Yu, J. Gao, J. C. Hummelen, F. Wudl, and A. J. Heeger. Science 270 (1995) 1789.
[3] K. Harada, A. G. Werner, M. Pfeiffer, C. J. Bloom, C. M. Elliott, and K. Leo. Phys. Rev. Lett. 94 (2005) 036601.
[4] K. Walzer, B. Maennig, M. Pfeiffer, and K. Leo. Chem. Rev. 107 (2007) 1233.
[5] J. Blochwitz, M. Pfeiffer, T. Fritz, K. Leo, D.M. Alloway, P.A. Lee, N.R. Armstrong. Org. Electron. 2 (2001) 97.
[6] K. Harada, M. Riede, K. Leo, O. R. Hild, C. M. Elliott. Phys. Rev. B 77 (2008) 195212.
[7] C. W. Law, K. M. Lau, M. K. Fung, M. Y. Chan, F. L. Wong, C. S. Lee, and S. T. Leeb. Appl. Phys. Lett. 89 (2006) 133511.
[8] Th. Singh, F. Meghdadi, S. Guenes, N. Marjanovic, G. Horowitz, P. Lang, S. Bauer, and N. S. Sariciftci. Adv. Mater. 17 (2005) 2315.
[9] H. Ishii, K. Sugiyama, E. Ito, and K. Seki. Adv. Mater. 11 (1999) 605.
[10] I.G. Hill, J. Schwartz, A. Kahn, Org. Electron. 1 (2000) 5.
[11] X. Crispin, V. Geskin, A. Crispin, J. Cornil, R. Lazzaroni, W. R. Salaneck, and J.-L. Bredas. J. Am. Chem. Soc. 124 (2002) 8131.
[12] W. Gao and A. Kahn. Appl. Phys. Lett., 79 (2001) 10.
[13] N. Koch, A. Elschner, R.L. Johnson, and J.P. Rabe. Appl. Surf. Sci. 244 (2005) 593.
[14] F. Amy, C. Chan, and A. Kahn. Org. Electr. 6 (2005) 85.
[15] H. Ding and Y. Gao. Appl. Surf. Sci. 252 (2006) 3943.
[16] H. Ding and Y. Gao. J. Appl. Phys. 102 (2007) 043703.
[17] C. Chan, E.-G. Kim, J.-L. Bredas, and A. Kahn. Adv. Funct. Mater. 16 (2006) 831.
[18] S. Olthof and W. Tress, R. Meerheim, B. Luessem, and K. Leo. J. Appl. Phys. 106 (2009) 103711.
[19] N. Koch and A. Vollmer. Appl. Phys. Lett. 89 (2006) 162107.
[20] I. G. Hill, A. Rajagopal, and A. Kahn Y. Hu. Appl. Phys. Lett. 73 (1998) 662.
[21] K. Ihm, H.-E. Heo, S. Chung, J.-R. Ahn, J. Kim, and T. Kang. Appl. Phys. Lett. 90 (2007) 242111.

Mater. Res. Soc. Symp. Proc. Vol. 1270 © 2010 Materials Research Society 1270-II09-53

The Ultrafast Dynamics of Electronic Excitations in Pentacene Thin Films

Henning Marciniak[1], Bert Nickel[2], and Stefan Lochbrunner[1]
[1]Institut für Physik, Universität Rostock, Universitätsplatz 3, 18055 Rostock, Germany
[2]Fakultät für Physik und CeNS, Ludwig-Maximilians-Universität,
Geschwister-Scholl-Platz 1, 80539 München, Germany

ABSTRACT

Pentacene films which are model systems for organic electronics exhibit contrary to the pentacene monomer extremely little photoluminescence. This points to an ultrafast relaxation mechanism. We apply pump-probe absorption spectroscopy with a time resolution of 30 fs to microcrystalline pentacene films. It is found that the primarily excited Frenkel excitons decay within 70 fs to a non fluorescing singlet species. Contrary to expectations fission into triplets is only a minor channel. We propose that the ultrafast relaxation of the photoexcited excitons leads to a species very similar to excimers. The radiative transition of the excimer to the ground state is electric dipole forbidden due to symmetry reasons. The excimer formation provides therefore an efficient mechanism to turn off the emission.

INTRODUCTION

Pentacene is one of the most promising candidates for applications in organic electronics like transistors because of its extraordinary high hole mobility in the crystalline phase of more than 1 cm^2/Vs [1]. Therefore pentacene serves as model compound to understand the electronic properties of organic crystals. In optoelectronic devices the behavior of electronic excitations like excitons is crucial for the performance. While the pentacene monomer fluoresces with a high quantum yield, pentacene films show only an extremely low photoluminescence pointing to the relevance of ultrafast relaxation processes. To investigate this dynamics, we apply pump-probe absorption spectroscopy with a time resolution of 30 fs to microcrystalline pentacene films.

EXPERIMENT

Microcrystalline thin film samples are prepared by vacuum deposition of pentacene on a transparent polymer substrate (TOPAS; thermoplastic olefin polymer of amorphous structure). The film consists of closely packed crystalline grains with a diameter of about 1 μm and a height of roughly 30 monolayers [2]. The samples are kept under nitrogen for storage and during the optical experiments. In the pump probe measurements the samples are excited by 30 fs long pulses which are generated with a noncollinearly phase matched optical parametric amplifier (NOPA) pumped by a 1 kHz regenerative Ti:sapphire amplifier system (CPA 2001; Clark MXR). The absorption changes are probed over the whole visible spectral range with a whitelight continuum generated in a 3 mm thick sapphire substrate. The photoexcitation is performed at the lowest absorption band with an excitation wavelength of 669 nm.

RESULTS AND DISCUSSION

Transient spectra

Figure 1 shows the steady state absorption spectrum of the investigated pentacene film and transient absorption spectra measured with different polarization geometries 8 ps after photo excitation. They are dominated by the bleach of the ground state absorption. The positive absorption change around 630 nm in the case of parallel polarizations demonstrates that excited state absorption (ESA) contributes too. The dependence on the polarization reflects the crystalline order and Davydov splitting. The pentacene molecules are standing almost perpendicular on the substrate and form a herringbone structure. The transition dipole of the molecular S_1-S_0 transition is along the short molecular axis. Both resulting Davydov components are parallel to the surface but almost orthogonal to each other [3,4]. The lower component contributes particularly strong to the bleach for parallel polarizations while the upper one for perpendicular polarizations. The orientation of the herringbone structure varies from grain to grain. If the excitons generated by the excitation would hop between the grains the polarization dependence should disappear with time. This is not observed indicating that the excitons are not mobile enough to change between grains. However, they are not immobile. This was shown by studies with varying excitation energies. They revealed pronounced exciton-exciton annihilation. From the analysis of the annihilation dynamics we found a diffusion coefficient for the excitons of 5×10^{-4} cm^2/s [5].

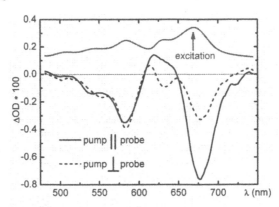

Figure 1. The steady state absorption spectrum (grey line) of the pentacene films and transient absorption spectra 8 ps after excitation at 669 nm. The transient spectra are recorded with the probe polarization parallel (black solid line) and perpendicular (broken line) to the pump polarization.

Ultrafast electronic dynamics

The intrinsic lifetime of the excitons is too long to account for the ultralow fluorescence yield of crystalline pentacene. Therefore we investigated in more detail the spectral region around the red wing of the first absorption band where stimulated emission is expected. Figure 2 shows the spectral evolution in this region during the first 250 fs and kinetic traces measured at 635 nm and 676 nm.

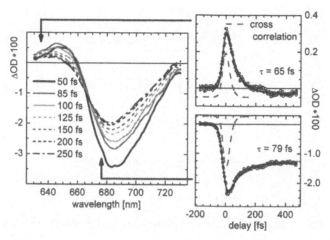

Figure 2. Transient spectra under normal incidence during the first 250 fs in the region of the first absorption band (left) and time traces (circles) measured with 25 fs probe pulses centered at 635 nm and 676 nm (right). Monoexponential decays (solid lines) are convoluted with the crosscorrelation (dash-dotted lines, vertically shifted for better visibility) and fitted to the data.

At the low energy side of the first steady state absorption band an originally very strong and negative signal is observed which decays rapidly to the transient spectrum known from the measurements taken at longer delay times. Since in this spectral region stimulated emission should occur we interpret the observed dynamics as ultrafast fluorescence decay. From the kinetic traces a decay time of 70 fs is derived [6]. This is in agreement with the low fluorescence yield.

Indications for such an ultrafast process have been found before and were interpreted as fission of the primarily generated Frenkel excitons into triplet excitons [3]. Since per each singlet exciton two triplet excitons are generated the total spin is conserved. In tetracene crystals this process occurs on the sub-nanosecond time scale and is associated with a thermal activation energy of 0.21 eV [7]. As shown in Figure 3, the triplet energy of the pentacene monomer is slightly less than half of the first electronically excited singlet state. Extrapolating the fission rate of tetracene to the energetics of pentacene predicts a sub-100 fs decay for the singlet excitons. Consequently the ESA occurring in the transient spectra around 630 nm was attributed to triplet excitons [3].

However, this assignment does not match the triplet spectrum of pentacene monomers in solution [8]. There the energy difference between triplet and singlet absorption is about 10 times larger than the energy difference between the ground state bleach and the ESA in the transient spectra of the microcrystalline films.

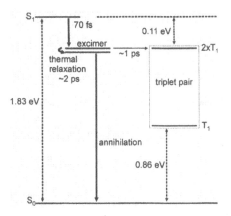

Figure 3. Energy diagram for the species involved in the relaxation of optically excited excitons (S_1). T_1 indicates the energy level of a triplet exciton.

It is also problematic to understand the dynamics on longer time scales in terms of triplet excitons. As pointed out above this dynamics is dominantly due to exciton-exciton annihilation [5]. Since the energy of two triplet excitons should be below the energy of one singlet exciton two encountering triplets cannot create a highly excited state and it is unlikely that they annihilate efficiently. In summary, a more specific triplet signature is needed to get a clear picture of the electronic dynamics in crystalline pentacene films.

Delayed triplet formation

A delicate balance determines the strength of the triplet absorption in pentacene films. The transition dipole of the lowest triplet-triplet absorption band is quite big and the absorption cross section is more than 10 times larger than that of the lowest singlet absorption [8]. The dipole is aligned along the long molecular axis and the pentacene molecules in crystalline films with their typical herringbone structure are standing almost upright on the substrate [9]. At normal incidence of pump and probe pulses the electric fields are almost perpendicular to the triplet transition dipole and absorption signatures from triplet excitons are strongly suppressed.

To obtain triplet specific signatures, measurements were performed with an angle of incidence of 65° for the laser beams [5]. The results are shown in Figure 4. The transient spectra (grey line) exhibit similar features as the spectra measured at normal incidence (black line) but an additional ESA band is clearly observed around 540 nm which does not appear at normal incidence. The spectral position is slightly red shifted by 0.16 eV with respect to the triplet absorption of the monomer. Therefore we assign this signature to triplet excitons while the ESA around 630 nm results from a species with singlet character. Taking the large triplet absorption cross section into account, the moderate strength of the band indicates that the triplet population is a minority.

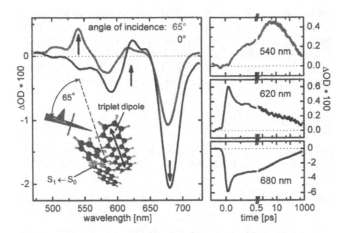

Figure 4. Left: Transient spectra of a thin pentacene film 8 ps after excitation at 669 nm with parallel polarized pump and probe beam for different angles of incidence. Right: Time scans at different probe wavelengths taken under the same conditions. Grey curves denote measurements with an angle of incidence of 65° and black curves with normal incidence.

The kinetic trace measured at 540 nm (see Fig. 4) shows an absorption rise on the picosecond timescale and not within 70 fs contrary to the traces at 620 nm and 680 nm. This indicates that the triplet excitons are formed in a subsequent process after the first ultrafast relaxation step. The corresponding time constant does not significantly contribute to the dynamics in other spectral regions. This is a further indication that the fission into triplets is a minor relaxation channel. At longer delay times a signal decay is observed at 540 nm. However, we observe a corresponding signal rise at 650 nm (not shown) indicating that the triplet excitons are trapped or experience another relaxation process but do not decay within the first nanosecond.

CONCLUSIONS

The presented measurements show that in microcrystalline pentacene films the primarily excited Frenkel excitons decay within 70 fs to a non fluorescing singlet species. Contrary to expectations fission into triplets is only a minor channel. We propose that the ultrafast relaxation of the photoexcited excitons leads to a species very similar to excimers. Neighboring pentacene molecules form dimers which are bound stronger in the electronically excited state than in the ground state. The relaxation time probably reflects geometric rearrangements like a rotation around the long molecular axis and a reduction of the intermolecular distance to increase the overlap of the π-orbitales and the interaction strength. The resulting gain in energy should be in the order of a few 100 meV (see Fig. 3). In this case the excimer energy is smaller than two times the triplet energy and fission into triplets should be a thermally activated process. This results in a picosecond rate for the triplet formation directly after the excimers are generated and as long as the released energy has not yet dissipated. After this energy is distributed over a larger sample region the local temperature has dropped so far that the fission rate becomes irrelevant. Accordingly, only a minor fraction of the excitons is transformed into triplets. Like in H-aggregates the radiative transition of the excimer to the ground state is electric dipole forbidden due to symmetry reasons. In this way the excimer formation provides an efficient mechanism to turn off the emission within 70 fs as it is observed in the experiment.

ACKNOWLEDGMENTS

We thank Matthias Fiebig and Stefan Schiefer for sample preparation and Florian Selmeier for support in the spectroscopic experiments.

REFERENCES

1. C. D. Dimitrakopoulos, S. Purushothaman, J. Kymissis, A. Callegari, and J. M. Shaw, Science **283**, 822 (1999).
2. B. Nickel, M. Fiebig, S. Schiefer, M. Göllner, M. Huth, C. Erlen, and P. Lugli, Phys. Status Solidi A **205**, 526 (2008).
3. C. Jundt, G. Klein, B. Sipp, J. Le Moigne, M. Joucla, and A. A. Villaeys, Chem. Phys. Lett. **241**, 84 (1995).
4. K. O. Lee and T. T. Gan, Chem. Phys. Lett. **51**, 120 (1977).
5. H. Marciniak, I. Pugliesi, B. Nickel, and S. Lochbrunner, Phys. Rev. B, **79**, 235318 (2009).
6. H. Marciniak, M. Fiebig, M. Huth, S. Schiefer, B. Nickel, F. Selmaier, and S. Lochbrunner, Phys. Rev. Lett. **99**, 176402 (2007).
7. M. Pope, N. E. Geacintov, and F. Vogel, Mol. Cryst. Liq. Cryst. **6**, 83 (1969).
8. C. Hellner, L. Lindqvist, and P. C. Rodberge, J. Chem. Soc., Faraday Trans. 1 **68**, 1928 (1972).
9. S. Schiefer, M. Huth, A. Dobrinevski, and B. Nickel, J. Am. Chem. Soc. **129**, 10316 (2007).

Mater. Res. Soc. Symp. Proc. Vol. 1270 © 2010 Materials Research Society   1270-II09-79

Structure analysis of solution-crystallized 2,7-dioctylbenzothieno[3,2-b]benzothiophene thin films in very high-mobility transistors

J. Soeda[1], M. Yamagishi[1], Y. Hirose[1], T. Uemura[1,2], A. Nakao[3], Y. Nakazawa[1], S. Shinamura[4], K. Takimiya[3], and J. Takeya[1,2]
[1]Graduate School of Science, Osaka University, Toyonaka 560-0043, Japan.
[2]ISIR, Osaka University, Ibaraki 567-0047, Japan.
[3]High Energy Accelerator Research Organization, Tsukuba, Ibaraki 305-0801, Japan.
[4]Graduate School of Engineering, Hiroshima University, Higashi-Hiroshima 739-8527, Japan.

ABSTRACT

We studied crystalline analysis on recently developed crystalline films grown from a solution in an oriented way on a substrate. The film is characterized as very high mobility values reaching 5 cm^2/Vs for devices with solution-crystallized 2,7-dioctylbenzoselenopheno[3,2-b]benzoselenophene (C_8-BTBT) thin films. High-power transmission X-ray scattering is employed to examine the crystallinity of the films, to identify whether or not the structure is same as the bulk structure, and determine the direction of the orientational crystal growth.

INTRODUCTION

Organic field effect transistors (OFETs) have been considerably interested due to their ability of application to flexible, large area and low-cost electrical switching components. Although it is mostly argued to use them for relatively slow devices, further applications become possible with the achievement of higher mobility above ~10 cm^2/Vs, which is demonstrated in single-crystal OFETs, in practically mass-producible transistors. Unfortunately, the conventional single-crystal OFETs have not been suited to fabricated large-scale devices. Very recently we developed a method to grow a crystalline film from a solution in an oriented way on a substrate and reported considerable mobility values exceeding a few cm^2/Vs for devices with solution-crystallized 2,7-dioctylbenzoselenopheno[3,2-b]benzoselenophene (C_8-BTBT) thin films [1]. Apparently, the film consists of crystalline domains spread to the whole channel length, so that molecular steps and terraces are observed in atomic-force-microscope view. In this study, we performed high-energy transmission X-ray diffraction measurement of the solution-crystallized C_8-BTBT films on typically 2-µm-thick parylene substrates. The measurement was configured to elucidate crystallinity of the film, favored direction of the crystal growth, relationship between direction of the crystal and that of favorable carrier transport.

Obvious Bragg peaks emerged corresponding to the components in the conducting a-b plane consistently to the reported crystal structure of the material. The peak patterns indicate that the C_8-BTBT films consist of either one or a few crystal domains, within the sub-millimeter spot of the irradiated X-ray. It turned out that the c-axis is vertical to the substrate and that the a axis are inclined by approximately 30 degrees to the direction of crystal growth. The growth direction nearly corresponds to that of the highest average transfer integral. Note that the high-mobility transistor performance is measured in the same direction. We conclude that the high performance in the solution-processed C_8-BTBT thin-film transistors are originated from extremely ordered molecular stacking due to the crystallization.

EXPERIMENT

C_8-BTBT derivatives are synthesized by one of the coauthors with excellent TFT performances both in vacuum-deposited and spin-coated thin films [2,3]. The molecules include alkyl chains in addition to the BTBT main structure, as shown in Figure 1, so that the chains give functions of enhancing solubility and "fastening" adjacent molecules through mutual attractive interactions. Therefore, the crystalline analysis indicates that molecular distances are shorter for molecules with longer alkyl chains, resulting in the tendency of larger mobility in the BTBT TFTs with longer alkyl chains [3]. The attractive interaction also facilitates rapid crystal growth so that multi-domain thin-films are easily fabricated on the substrates. The highest mobility reported so far was 2.3 cm^2/Vs for solution-processed films of C_{13}-BTBT, although the reproducibility was rather poor [2]. In the present experiment, we use C_8-BTBT, which showed the highest mobility of 1.8 cm^2/Vs.

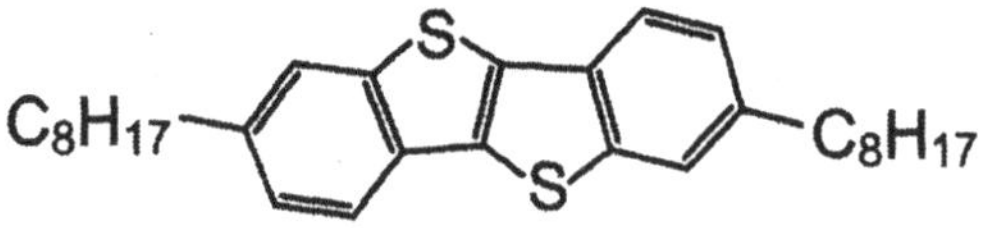

Figure 1. Molecular structure of C_8-BTBT.

Figure 2 schematically illustrates our present method of the oriented growth of the crystalline thin films on SiO_2 / doped Si substrates. After the surface treatment of vapor depositing decyltriethoxysilane (DTS), a 0.4wt% solution of C_8-BTBT is prepared with a solvent of heptane and a droplet is sustained at an edge of a structure on an inclined substrate, so that the crystalline domain grows in the direction of the inclination through evaporation of the solvent. The structure to support the droplet can be a small piece of a silicon wafer, for example, and can be removed after the growth of the crystalline film. In order to thoroughly remove the solvent, we dried it in a vacuum oven for typically 5 h at 50 C. Source and drain electrodes are then evaporated on the film, so that the channel is parallel to the direction of inclination, i.e., the growth direction. The length and width of the channel are 0.1 and 1.5 mm, respectively. The doped-Si layer acts as a gate electrode, so that electric field is applied to the 500-nm-thick SiO_2 whose dielectric constant is approximately 3.9. Figure 3 shows optical micrograph of the device, where a very homogeneous crystal-like surface is already visible.

Figures 4(a) and 4(b) show more microscopic surface morphology of the resultant film of C_8-BTBT taken by an atomic-force microscope. Over the range of micormeter-scale terraces, the surface is molecularly flat and the steps are of one additional molecular layer with the hight of the C_8-BTBT molecule, The excellent flatness and the well-defined molecular steps indicate that highly oriented crystalline films are grown on the SiO_2 dielectric surface.

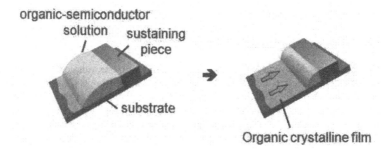

Figure 2. A schematic illustration of the method of orientational crystallization of organic semiconsutor thin films from solutions.

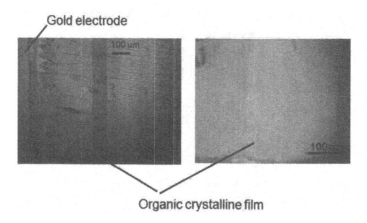

Figure 3. Optical views of the solution grown crystalline films of C₈-BTBT.

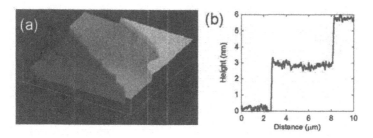

Figure 4. Atomic-force-microscope views of the solution grown crystalline films of C₈-BTBT.

RESULTS AND DISCUSSION

Analysis of High-power X-ray diffraction experiments

We performed high-power diffraction experiments with transmission X-ray configuration. First, the incident direction is set nearly parallel to the substrate to confirm the molecular alignment in the c-axis direction, which is indicated from the result of the AFM measurement. Figure 5 shows the imaging-plate picture and the result of ϕ-scan measurement, showing the X-ray intensity as a function of diffraction angle. The well-defined spots in the imaging plate and the sharp peaks in the ϕ-scan result demonstrate the c-axis ordering is indeed very high.

More interestingly, Figure 6 shows the results of similar experiments to detect the molecular ordering in the ab-plane direction. The ϕ-scan means the angle dependence in the ab-plane this time. Again obvious Raue spots appeared in the imaging plate and the peaks in the ϕ-scan measurements are very sharp, evidencing that the thin film in the regime of submillimeters is of real single crystal. The size is larger than that of the channel length.

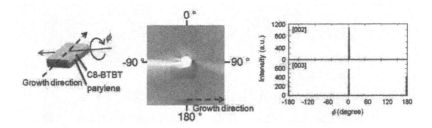

Figure 5. X-ray diffraction spots when the incident X-ray is nearly parallel to the substrate.

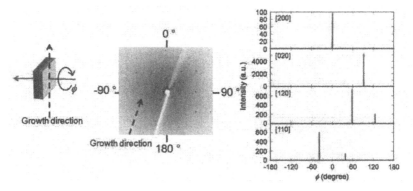

Figure 6. X-ray diffraction spots when the incident X-ray is nearly perpendicular to the substrate.

210

We also note that the positions of the spots are all consistent to the crystal structure gicen for the bulk [2], which means the structure is most likely same as that in the present thin films.

Performances of the solution grown C$_8$-BTBT crystalline thin-film transistors

Plotted in Figure 7 is the transfer characteristics of the C$_8$-BTBT TFT in the saturation region. The slope of the plot gives mobility as high as 5 cm^2/Vs. The subthreshold swing is as low as 3 V/decade. We note that most of good-looking devices prepared by the same method had mobilities in the range of 3.5-5 cm^2/Vs, though they can be less than 2 cm^2/Vs in devices with significant crystalline breaks visible under optical microscope. The yield could be further improved by reducing the device size and drying the semiconductor layer more gently. The on-off ratio exceeds 10^6 for all the devices.

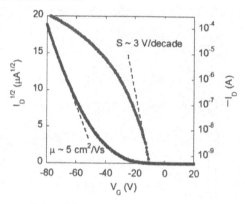

Figure 7. Transfer characteristics of the solution-crystallized C8-BTBT transistor.

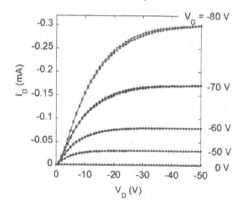

Figure 8. Output characteristics of the solution-crystallized C8-BTBT transistor.

Figure 8 shows the output characteristics of the same device. Well-saturated behavior with negligible hysteresis ensures that injection and transport of accumulated carriers reproduce the typical properties of transistor function described by the standard gradual-channel approximation. A room to improve the performance of the devices is in a non-Ohmic current-voltage characteristics at low drain voltages. The results indicate the presence of injection barriers for holes due to the relatively deep highest occupied molecular orbital (HOMO) level of BTBT [2]. The mobility of 5 cm^2/Vs is achieved in the saturation regime with relatively high drain and gate voltages, whereas only about 2 cm^2/Vs is estimated as a linear mobility when the drain voltage is 10 V, for example.

Finally, we address that the direction of the crystal growth differs from both crystallographic a- and b-axes, as shown in Figure 9. The direction of the highest molecular overlap is that of a-axis. Since the devices are fabricated so that the channel is parallel to the growth direction, there would be a room to improve their performance by optimizing the channel direction.

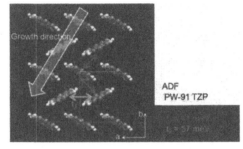

Figure 9. Crystal-growth direction and crystallographic axes in the C$_8$-BTBT solution-crystallized films.

CONCLUSIONS

We have demonstrated that high-performance organic single-crystal transistors can be fabricated by an easy process based on orientational crystal growth on substrates. Mobility of the C$_8$-BTBT crystalline film reaches 5 cm^2/Vs. The high-power X-ray diffraction revealed that the channel region is really of single crystal.

ACKNOWLEDGMENTS

BTBT is supplied from Nihon Kayaku Co., ltd.. This work was financially supported by Industrial Technology Research Grant Program from NEDO, Japan, and by a Grant-in-Aid for Scientific Research (Nos. 21108514 and 22245032) from MEXT, Japan.

REFERENCES
1. T. Uemura, Y. Hirose, M. Uno, K. Takimiya, and J. Takeya, *Appl. Phys. Exp.* **2**, 111501(2009).
2. H. Ebata, T. Izawa, E. Miyazaki, K. Takimiya, M. Ikeda, H. Kuwabara, and T. Yui, *J. Am. Chem. Soc.* **129**, 15732 (2007).

3. T. Izawa, E. Miyazaki, and K. Takimiya, *Adv. Mater.* **20**, 3388 (2008).

Interfaces and Morphology

Mater. Res. Soc. Symp. Proc. Vol. 1270 © 2010 Materials Research Society   1270-II11-10

Laser printing of organic insulator and semiconductor based thin films for transistor applications

Ludovic Rapp[2], Sébastien Nénon[1], Abdou Karim Diallo[1], Anne Patricia Alloncle[2], Philippe Delaporte[2], Christine Videlot-Ackermann[1*] and Frédéric Fages[1]

[1] CINaM (Centre Interdisciplinaire de Nanoscience de Marseille) – UPR 3118 CNRS- Campus de Luminy, Case 913, 13288 Marseille Cedex 09, France

[2] Laboratoire LP3 (Lasers, Plasma et Procédés Photoniques) - UMR 6182 CNRS - Université de la Méditerranée - Campus de Luminy C917, 13288 Marseille Cedex 09, France

ABSTRACT

This paper presents a pulsed-laser printing (PLP) process applied to an insulating polymer and a semiconducting oligomer with the goal to fabricate organic thin film transistors (OTFTs). The laser-induced forward transfer (LIFT) technique has been used as a spatially-resolved laser deposition method. All materials have been transferred from a donor substrate onto a receiver substrate upon laser pulses in the picosecond regime. The broad nature of transferred patterns and the efficiency of the LIFT confirm the important potential of a laser printing technique in the development of the plastic microelectronics. Electrical characterizations in OTFT configurations demonstrated that transistors are fully operative.

INTRODUCTION

Organic thin film transistors (OTFTs) have been of great interests for decades, with their applications as driving and switching elements in electronic and optical devices [1]. They offer unique opportunities in low-cost microelectronics as the fabrication of devices on flexible substrates. The success of their development depends on the ability of scientists and engineers to print thin-film transistors with performances compatible with the targeted applications. The main disadvantages of approaches used actually, as roll to roll and inkjet printing techniques, for printing thin film components are the use of soluble materials, and they often require a step of annealing which may prevent the use of some low-cost flexible substrates. Indeed, the development of facile and cheap fabrication processes allowing the deposition of a large set of materials (organic and inorganic) with micrometer size resolution represents a task of major importance. Laser-based processes [2, 3] offer versatile alternatives for the deposition of thin films in organic devices operating on flexible supports where usual techniques cannot be used due to a lack of solubility or in the case of complex device architectures fabrication.

The laser-induced forward transfer (LIFT) technique [4] is a promising alternative for fabrication of organic and metallic electronic components. This single step, direct printing technique offers the ability to make surface micro patterning or localized deposition of material. It can be applied to sensitive materials without altering their properties but it also allows to direct-write multilayer systems in a solvent-free single step, without requiring any shadowing mask or vacuum installation.

In previous work, we could demonstrate the pulsed-laser printing of different kind of materials, as an insulator poly(methyl methacrylate) (PMMA) [5] and an organic semiconductor copper phthalocyanine (CuPc) [6]. This paper presents a more comparative study on the optical absorption properties of such materials as well as OTFT's preparation and performances. PMMA and CuPc were chosen for their different physical and electrical properties. The challenges are to obtain a high spatial resolution of the transferred materials along with their initial properties. For each material, OTFTs were realized and their performances measured.

EXPERIMENT

The LIFT technique consists in removing a piece of a thin layer previously deposited on a transparent substrate (donor substrate) and transferring it on another substrate (receiver substrate) using a pulsed laser [7, 8]. The donor substrate must be transparent at the laser's wavelength. UV-transparent quartz suprasil (Hellma-France) were used. Figure 1 shows the principle of the pulsed-laser printing process. The film to be printed is irradiated through the transparent substrate, and the light-matter interaction, which takes place at the interface, generates a strong increase of the local pressure. As a result, a small piece of the thin film is ejected and deposited onto the receiver substrate. The laser is a 50 ps Nd:YAG source (continuum leopard SS-10-SV) used on its third harmonic (355 nm) and the experiments are performed at room temperature and ambient pressure. The fluence is controlled with calibrated polarization devices and a square mask is used to select a homogeneous part of the beam to be imaged on the donor thin layer with a converging lens. A mechanical shutter is used to select one pulse. Donor and receiver substrates are perpendicular to the laser beam axis, and spacers are used to keep the distance between them below ten micrometers. The precise positioning of the sample is obtained by micrometric translation stages and controlled by imaging the spot with a CCD camera.

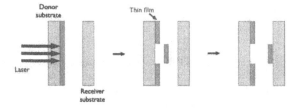

Figure 1. Schematic representation of the pulsed-laser printing process.

The poly(methyl methacrylate) (PMMA), a transparent thermoplastic, can be used as a dielectric layer deposited by spin-coated with a high resistivity (> 2 10^{15} Ω.cm), a low dielectric constant (ε = 2,6 at 1 MHz, ε = 3,9 at 60 Hz) and a low roughness. Additionally, PMMA presents great advantages as component in transparent flexible devices with a high mechanical resistivity, a low glass transition temperature (inferior at 120°C) and a high transparency in the visible range. As organic compounds, phthalocyanines are today clearly regarded as optical materials and organic dye layers. The electrical and optical properties of the Pc are determined by a central metal ion and side groups. Copper phthalocyanines with the Cu as the central metal ion, have excellent chemical and thermal stability in environmental condition. The copper phthalocyanine bearing hydrogen atoms as side groups ($H_{16}CuPc$ or CuPc) shows p-type semiconducting nature [9]. It has been known that phthalocyanines with high thermal and chemical stability represent one of the most promising candidates for modern opto-electronic devices such as optical recording, organic light-emitting diodes, gas sensors, solar cells and OTFTs [10-15]. CuPc and

PMMA were purchased from Aldrich (France) with a sublimation grade (purity > 99%) for CuPc and an average Mw of 15 000 (GPC) for PMMA. The molecular structures of PMMA and CuPc are shown on Figure 3. UV-visible absorption spectra of both thin films were obtained on a Varian Cary 1E spectrophotometer.

Two different OTFT devices (Figure 2) have, depending on the nature of the LIFT-printing deposit, been achieved:

- *LIFT printing of PMMA:* The donor substrate was made by spin-coating PMMA solution onto cleaned suprasil substrate (5 s at 1000 rpm and then 20 s at 2000 rpm). The measured thickness of PMMA-based donor layer was 700 nm. The receiver substrates were already formed by operational Bottom Gate (BG) transistors (BG-OTFT @ SiO_2 on Figure 2c). Silicon (Si) wafers oxidized by a 300 nm thick layer of silicon dioxide (SiO_2) (C_i = 12 nF/cm^2) purchased from Vegatec (France) were used as gate and dielectric, respectively. CuPc was thermally evaporated under high vacuum (2×10^{-6} mbar) on substrate maintained at room temperature to a nominal thickness of 100 nm. The receiver substrates were completed by the thermal evaporation of 50 nm thick of gold source and drain electrodes in top contact configuration with specific channel width (W = 1.7 mm) and channel length (L = 65-650 µm) (Figure 2a). LIFT-printed deposits of PMMA were realized on top of CuPc layer at a fluence of 0.21 J/cm^2. The average thickness of deposits was 400 nm (C_i = 5.6 nF/cm²). In order to complete the Top Gate-OTFTs (TG-OTFTs @ PMMA on Figure 2c), aluminium (Al) was thermally evaporated under high vacuum to form the gate on the printed PMMA with a nominal thickness of 50 nm. Top Gate-OTFTs had defined with a channel width equal to the PMMA deposit length (W = 500 µm) with a channel length of L = 80 µm.

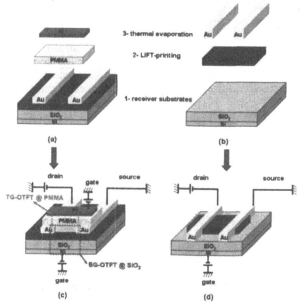

Figure 2. Schematic representation of OTFT devices using LIFT-printed deposits of PMMA (a, c) and CuPc (b, d).

219

- *LIFT printing of CuPc:* The donor substrate has been prepared by thermal evaporation of CuPc under vacuum (2×10^{-6} mbar at room temperature) on UV-transparent quartz suprasil to a nominal thickness of 100 nm. The average thickness of deposits was 80 nm. The most homogeneous pixels were printed at the low fluence of 0.10 J/cm². The receiver substrates are silicon (Si) wafers oxidized by a 300 nm thick layer of silicon dioxide (SiO_2) (from Vegatec, France) with $C_i = 12$ nF/cm². These materials served as gate and dielectric in transistors (Figure 2d). The receiver substrates are completed by the thermal evaporation of 50 nm thick of gold source and drain electrodes in top contact configuration with a channel width equal to the CuPc deposit length (W = 500 µm) together with a channel length L form 40 to 80 µm. (Figure 2b).

Current-Voltage characteristics were obtained with a Hewlett-Packard 4140B pico-amperemeter DC voltage source under Labview® environment. All the measurements were performed at room temperature under ambient atmosphere. The field effect mobility µ was extracted from the transfer characteristics in the saturation regime. The drain-source current ($I_{drain\text{-}source}$) in the saturation regime is governed by the equation:

$$(I_{drain\text{-}source})_{sat} = WC_i\mu / 2L \, (V_{gate}\text{-}V_t)^2 \tag{1}$$

where C_i is the capacitance per unit area of the gate insulator layer, V_{gate} is the gate voltage, V_t is the threshold voltage, and µ is the field-effect mobility.

DISCUSSION

The optical absorption spectra of PMMA and CuPc deposited onto quartz substrates were recorded in a UV-Visible range from 200 to 900 nm (Figure 3). Thin films of PMMA and CuPc were realized by spin-coating and by thermal evaporation under vacuum, respectively. CuPc presents two large bands between 200-450 and 550-850 nm. At the laser irradiation (355 nm), CuPc-based thin films are highly absorbent, while PMMA thin films are mostly transparent. However, laser energy can be deposited in PMMA by two photon absorption when ultrashort pulses and sufficient power density is used [16]. Then, this process can lead to the PMMA film ejection.

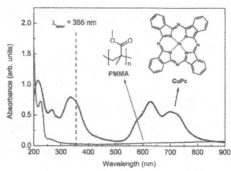

Figure 3. Absorption spectra of PMMA and CuPc thin films deposited on donor substrates.

Depending on such different optical absorption property of thin films to be LIFT printed at 355 nm, the fluence has to be optimized for each materials. The ablation of PMMA layer at a

220

low fluence, *i.e.* 0.10 J/cm², resulted in the transfer of only small fragments of the polymer. In increasing the fluence to 0.15 J/cm², the deposit was still partial. The most homogeneous transfer was obtained at 0.21 J/cm², as shown in Figure 4a. The high fluence used is directly correlated to the low absorbance of PMMA at 355 nm, requiring a higher initial energy to ablate the layer from the donor substrate. In the case of CuPc, LIFT printing occurred at a fluence of 0.10 J/cm². A single pulse leads to a homogeneous pixel of 500×500 μm² with well-defined edges (Figure 4b) with nevertheless some splashes all around the deposit.

Figure 4. Optical image of LIFT-printed deposits of PMMA (a) and CuPc (b).

Despite a transparency in the visible range, the high mechanical stability of PMMA as well as the high thickness of the layer (700 nm) makes possible its transfer by laser irradiation. In solution processes, the PMMA dielectric is often incompatible, as it is prepared itself in solution, with many organic semiconductors where orthogonal solvents can not be used. Laser transfer of an operational PMMA dielectric layer was realized in the present study by measuring the OTFT parameters of TG-OTFT @ PMMA (green dashed square in Figure 2a). Figure 5a shows the typical output characteristic of a fully operational TG-OTFT. Mobility values (μ) of 7.3-8.6×10⁻³ cm²/V.s and threshold voltages (V_t) between -1 and +55 V are obtained. As a comparison for OTFT devices based on vacuum evaporated CuPc active layer, (BG-OTFT @ SiO₂ in red dashed square in Figure 2a) present mobility values of 1.7-2×10⁻³ cm²/V.s and threshold voltages of (-1.8)-(+9) V. Despite an average mobility higher for TG-OTFT @ PMMA directly correlated to the dielectric nature, a large dispersion of V_t reveals a density of trap states to be filled before the carriers become mobile. In the case of LIFT-printed deposits of CuPc, an even lower mobility is obtained with 4×10⁻⁵ cm²/V.s. The threshold voltage is +6 V (Figure 5b). Indeed CuPc, as an organic small molecule, is more sensitive to the LIFT process. Nevertheless, such results confirm the potential of LIFT to print pixels of organic layers in transistor applications.

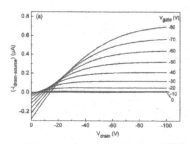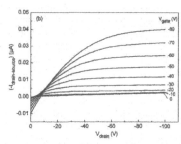

Figure 5. Output characteristic of an OTFT device using LIFT-printed deposits of PMMA (a) and CuPc (b).

CONCLUSIONS

In this study, functional organic transistors have been realized using a simple approach technology. The broad nature of transferred patterns (*i.e.* polymer *vs.* small molecule, transparent *vs.* dye, insulator *vs.* semiconductor) confirms the potential of LIFT as a very versatile technique. We have demonstrated the feasibility of a laser induced forward transfer of a transparent plastic as PMMA to form a dielectric layer on top of an existing organic active layer. Despite a transparency in the visible range, the high mechanical stability of PMMA makes possible its transfer by laser irradiation. An organic semiconductor, CuPc, with a lower mechanical stability is directly laser assisted transferred on a receiver substrate equipped by gold source-drain electrodes to form the active layer. Moreover, even if the performances obtained with this PLP process are lower than those already achieved with other deposition techniques, they are very promising regarding the simplicity of the LIFT process and its potential optimizations, in particular, the addition to the donor of an appropriate sacrificial layer [17]. Such layer traps the incident radiation thus protecting the material to be transferred, and also facilitates its ejection process.

ACKNOWLEDGMENTS

This work has been carried out within the ANR e-Plast program (ANR-06-BLAN-0295).

REFERENCES

1. C.D. Dimitrakopoulos and P.R.L. Malefant, *Adv. Mat.* **14**, 99 (2002).
2. G.B. Blanchet, C.R. Fincher and I. Malajovich, *J. Appl. Phys.* **94**, 6181 (2003).
3. G.B. Blanchet, Y-L. Loo, J.A. Rogers, F. Gao, C.R. Fincher, *Appl. Phys. Lett.* **82**, 463 (2003).
4. J. Bohandy, B. Kim and F. Adrian, *J. Appl. Phys.* **60**, 1538 (1986).
5. A.K. Diallo, L. Rapp, S. Nénon, A.P. Alloncle, C. Videlot-Ackermann, P. Delaporte and F. Fages, submitted.
6. L. Rapp, A.K. Diallo, A.P. Alloncle, C. Videlot-Ackermann, F. Fages and Ph. Delaporte, *Appl. Phys. Lett* **95**, 171109 (2009).
7. F.J. Adrian, J. Bohandy, B.F. Kim, A.N. Jette and P. Thompson, *J. Vac. Sci. Technol. B* **5** 1490 (1987).
8. J. Bohandy, B.F. Kim, F.J. Adrian and A.N. Jette, *J. Appl. Phys.* **63**, 1158 (1988).
9. Z. Bao, A.J. Lovinger and A. Dodabalapur, *Adv. Mater.* **9**, 42 (1997).
10. M. Ottmar, D. Hohnholz, A. Wedel and M. Hanack, *Synth. Met.* **105**, 145 (1999).
11. M. I. Newton, T. K. H. Starke, G. McHale, M. R. Willis, *Thin Solid Films* **360**, 10 (2000).
12. H. Deng, Z. Lu, Y. Shen, H. Mao, and H. Xu, *Chem. Phys.* **231**, 95 (1998).
13. J. Zhang, J. Wang, H. Wang, and D. Yan, *Appl. Phys. Lett.* **84**, 142 (2004).
14. K. Xiao, Y. Liu, G. Yu and D. Zhu, Appl. Phys. *A: Mater. Sci. Process.* **77**, 367 (2003).
15. R. Ye, M. Baba, K. Suzuki and K. Mori, *Thin Solid Films* **517**, 3001 (2009).
16. RE. Samad, LC. Courrol, AB. Lugao, AZ. de Freitas, ND. Vieira. *Radiation Phys. And Chem.* **79** (3), 355 (2010)
17. R. Fardel, M. Nagel, F. Nüesch, T. Lippert and A. Wokaun, *Appl. Phys. Lett.* **91**, 061103 (2007).

AUTHOR INDEX

SUBJECT INDEX

Printed in the United States
by Baker & Taylor Publisher Services